Certification Study Companion Series

The Apress Certification Study Companion Series offers guidance and hands-on practice to support technical and business professionals who are studying for an exam in the pursuit of an industry certification. Professionals worldwide seek to achieve certifications in order to advance in a career role, reinforce knowledge in a specific discipline, or to apply for or change jobs. This series focuses on the most widely taken certification exams in a given field. It is designed to be user friendly, tracking to topics as they appear in a given exam. Authors for this series are experts and instructors who not only possess a deep understanding of the content, but also have experience teaching the key concepts that support readers in the practical application of the skills learned in their day-to-day roles.

More information about this series at `https://link.springer.com/bookseries/17100`

Snowflake SnowPro™ Advanced Architect Certification Companion

Hands-on Preparation and Practice

Ruchi Soni

Apress®

Snowflake SnowPro™ Advanced Architect Certification Companion: Hands-on Preparation and Practice

Ruchi Soni
New Delhi, India

ISBN-13 (pbk): 978-1-4842-9261-7   ISBN-13 (electronic): 978-1-4842-9262-4
https://doi.org/10.1007/978-1-4842-9262-4

Managing Director, Apress Media LLC: Welmoed Spahr
Acquisitions Editor: Jonathan Gennick
Development Editor: Laura Berendson
Editorial Assistant: Shaul Elson
Copy Editor: Kim Wimpsett

Cover designed by eStudioCalamar

Distributed to the book trade worldwide by Springer Science+Business Media New York, 1 New York Plaza, Suite 4600, New York, NY 10004-1562, USA. Phone 1-800-SPRINGER, fax (201) 348-4505, e-mail orders-ny@springer-sbm.com, or visit www.springeronline.com. Apress Media, LLC is a California LLC and the sole member (owner) is Springer Science + Business Media Finance Inc (SSBM Finance Inc). SSBM Finance Inc is a **Delaware** corporation.

For information on translations, please e-mail booktranslations@springernature.com; for reprint, paperback, or audio rights, please e-mail bookpermissions@springernature.com.

Apress titles may be purchased in bulk for academic, corporate, or promotional use. eBook versions and licenses are also available for most titles. For more information, reference our Print and eBook Bulk Sales web page at www.apress.com/bulk-sales.

Any source code or other supplementary material referenced by the author in this book is available to readers on GitHub. For more detailed information, please visit www.apress.com/source-code.

Printed on acid-free paper

This book is dedicated to my family for loving and supporting me, especially my parents for believing in me and motivating me to pursue my dreams.
You guys are the best. ☺

Table of Contents

About the Author

Ruchi Soni is a technology leader and multicloud enterprise architect. She helps customers accelerate their digital transformation journey to the cloud and build next-generation apps on forward-looking platforms. She is a people person at heart and has deep industry knowledge and business expertise in architecting, building, and scaling future-ready platforms that are highly available and agile.

Ruchi's name is included in the Snowflake Data Superhero 2023 list (an elite group of only 73 Snowflake experts around the world). She leads the Snowflake Growth Market Practice in a Global Fortune 500 company and spearheads training and certifications, incubates the development of different accelerators, and operationalizes resilient migration factory solutions. She is a TOGAF 9 certified architect and has completed 15+ vendor certifications including SnowPro Core and Advanced Architect along with various cloud certifications. Outside of work, she is an avid reader and likes to travel and meditate.

About the Technical Reviewer

 **Adam Morton** is an experienced data leader and author in the field of data and analytics with a passion for delivering tangible business value. Over the past two decades Adam has accumulated a wealth of valuable, real-world experiences designing and implementing enterprise-wide data strategies and advanced data and analytics solutions as well as building high-performing data teams across the UK, Europe, and Australia.

Adam's continued commitment to the data and analytics community has seen him formally recognized as an international leader in his field when he was awarded a Global Talent Visa by the Australian government in 2019.

Today, Adam works in partnership with Intelligen Group, a Snowflake pureplay data and analytics consultancy based in Sydney, Australia. He is dedicated to helping his clients overcome challenges with data while extracting the most value from their data and analytics implementations.

He has also developed a signature training program that includes an intensive online curriculum, weekly live consulting Q&A calls with Adam, and supportive data and analytics professionals guiding members to become experts in Snowflake. If you're interested in finding out more, visit www.masteringsnowflake.com.

Acknowledgments

I am thankful to my mother for her continuous encouragement and motivation. Mom, words cannot express how grateful I feel toward you for everything you do for me!

I would also like to extend my heartfelt gratitude to my entire leadership for their guidance and support.

I extend my heartfelt thanks to my readers for believing in me.

Finally, I would like to thank my family and friends who helped me to finalize this book within a limited time frame.

Foreword

In the past, IT professionals developed specialization in specific technology areas largely through years and decades of work experience. However, the traditional route cannot produce specialized people in technology areas that are new and emerging. For emerging technology areas, in conjunction with hands-on experience, IT professionals must use all credible resources available to accelerate the specialization journey and establish their credibility through advanced certification programs. People who go through this journey ahead of others have the responsibility to contribute to the community by sharing and creating credible resources for others. I am glad that Ruchi did not just go through this specialization journey in Snowflake and become one of the first to get certified as SnowPro Advanced Architect but also realized the lack of credible resources to accelerate the specialization and Advanced Architect certification journey.

I have known Ruchi for years to be a very hands-on person, recognized in the data and AI industry for her deep technical skills in Snowflake. She has strong industry knowledge and business expertise in helping customers migrate to the cloud using modern data architecture frameworks. So, when she spoke to me after her certification and shared her plans to publish a book that would help others accelerate their specialization and Advanced Architect certification journey, I was really excited for her and the Snowflake community at large.

The SnowPro Advanced Architect Certification test is tough and designed for individuals with knowledge and skills on the Snowflake architecture. As a prerequisite, candidates should be SnowPro Core certified with good hands-on experience in Snowflake implementations. This certification tests the ability to

- Design data flow from source to consumption using the Snowflake data platform

- Design and deploy a data architecture that meets business, security, and compliance requirements

- Choose Snowflake and third-party tools to optimize architecture performance

- Design and deploy a shared dataset using the Snowflake Data Marketplace and Data Exchange

This book has been written in an easy-to-follow format keeping the Advanced Architect certification test in mind. It will help you

- Gain necessary knowledge to succeed in the test and apply the acquired practical skills to real-world Snowflake solutions

- Deep-dive into various topics that Snowflake specifically recommends for the SnowPro Advanced Architect Certification test

- Unleash the power of Snowflake in building a high-performance system with practical examples and in-depth concepts

- Identify gaps in your knowledge required for the test and narrow down focus areas of your study

- Optimize performance and costs associated with your use of the Snowflake data platform

- Broaden your skills as a data warehouse designer to cover the Snowflake ecosystem

Mukesh Chaudhary

Mukesh Chaudhary is a Data and AI Leader at Accenture. He comes with vast experience in migrating the data and ML landscape of enterprises to the cloud, modernizing data platforms, and unlocking the value of data through AI and ML. The views and opinions expressed in the Foreword are his own and not of Accenture.

Introduction

Master the intricacies of Snowflake and prepare for the SnowPro Advanced
Architect certification exam with this comprehensive study companion.
This book provides robust and effective study tools that help you prepare
for the exam and is designed for those who are interested in learning the
advanced features of Snowflake along with preparing for the SnowPro
Advanced Architect certification using task-oriented descriptions and
concrete end-to-end examples.

The purpose of this book is to provide a gentle and organized approach
through comprehensive coverage of every relevant topic on the SnowPro
Advanced Architect exam across the different domains defined for the
exam including accounts and security, snowflake architecture, data
engineering, and performance optimization. Reading this book and
reviewing the concepts in it helps you gain the necessary knowledge to
succeed on the exam.

This study guide includes the following:

- Comprehensive understanding of Snowflake's unique
  shared data, multicluster architecture

- Guidance on loading structured and semistructured
  data into Snowflake

- Understanding various client drivers available to
  connect with Snowflake for data loading and unloading
  including a deep dive into Kafka connectors

- Resource optimization and performance management
  through clustering keys, query profiles, tuning,
  materialized views, and warehouse

- Different options and best practices for data loading
  and unloading

- Understanding of the different Snowflake data-sharing
  options including secure views

- Deep dive into storage and data protection options
  including Time Travel, data replication, and failover
  and cloning

You'll also be well-positioned to apply your newly acquired practical skills to real-world Snowflake solutions. Your result from reading this book will be a deep understanding of Snowflake that helps in taking full advantage of Snowflake's architecture to deliver value analytics insight into your business.

CHAPTER 1

Exam Overview

Congratulations, readers, for taking your first step towards SnowPro Advanced: Architect certification. I hope that this study guide will be informative and helpful as you prepare for the SnowPro Advanced Architect certification exam. In this chapter, we will discuss what to expect from the exam, how the exam is structured, and how to interpret questions.

Why Certify?

Certification is a strong indicator to everyone that you have the required knowledge and skills for the job. Since competition for every job nowadays is high, certification works as a stamping authority confirming that you have the required knowledge and skills. In fact, in some organizations, certification is mandatory for certain jobs. It also helps individuals to advance in their careers and establish professional credibility.

Snowflake is one of the fastest-growing technologies in the data space. Given that this is a relatively new technology, there are not many experts on the ground. This challenge of increased demand with limited experts means there are thousands of jobs for Snowflake experts.

© Ruchi Soni 2023
R. Soni, *Snowflake SnowPro™ Advanced Architect Certification Companion*,
Certification Study Companion Series, https://doi.org/10.1007/978-1-4842-9262-4_1

Snowflake has recently launched advanced certifications that can really help you stand out in the data community. In addition to these work benefits, when you successfully pass an exam, you will receive a digital credential validating your Snowflake skills in the role, in addition to extending the expiration of your existing SnowPro Core certification (as specified in the Snowflake portal).

Get Started

This exam validates your knowledge of advanced concepts of Snowflake and your ability to create architectures using Snowflake services. The questions on the exam are structured to understand a candidate's expertise in the following aspects:

- Architect and create a data pipeline using Snowflake services
- Create a Snowflake architecture with the required security and governance
- Understand Snowflake services and related tools for optimized user performance
- Understand how data sharing works in Snowflake and the technologies used

Prerequisites

Passing the SnowPro Core certification exam is a prerequisite for taking any advanced certification exam. This is because the SnowPro Core certification exam tests your base expertise of implementing and migrating data to Snowflake and covers all the basic concepts that are needed to create a solid foundation, as shown in Figure 1-1.

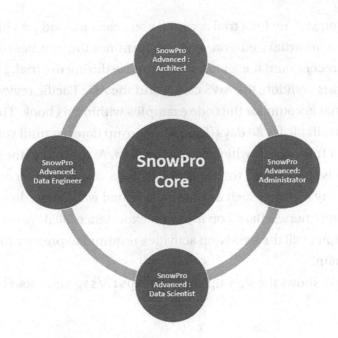

Figure 1-1. *Snowflake certification*

Since the ARA-C01 exam also has scenario-based questions, it is preferred if candidates have good hands-on experience (at least 2 years) working in Snowflake technology as an architect with sufficient knowledge of SQL and SQL analytics and experience in building out a complex ETL/ELT pipeline, implementing security and compliance requirements, and working with different data modeling techniques.

Create a Trial Account

To get sufficient hands-on experience to answer scenario-based questions on the exam, it is important to sign up for a free trial account. To create a trial account, you need only a valid email address; no payment information or other qualifying information is required.

When you sign up for a trial account, you select a cloud provider, a region, and a Snowflake edition, which determines the number of free credits you receive and the features you can use during the trial. I have used Enterprise edition, the AWS cloud, and the Asia Pacific region to create my trial account for the code examples within this book. This trial account is available for 30 days (from the sign-up date) or until you've used all your free credits, whichever occurs first. At the end of the trial, the account is suspended. You can still log into a suspended account but cannot use any features, such as running a virtual warehouse, loading data, or performing queries. Based on my exam experience, 30 days is sufficient time to complete all the hands-on activities required to prepare for the ARA-C01 exam.

Figure 1-2 shows the sign-up page at `https://signup.snowflake.com`.

Figure 1-2. *Snowflake trial account*

Exam Cost and Validity

The SnowPro Advanced: Architect certification exam currently costs $375, and the certification is valid for 2 years.

Exam Registration

Snowflake's certification exams are delivered through Pearson VUE and can be taken at any of the 1,000+ testing centers located globally or can be taken remotely in your home with a virtual proctor. As a participant of the SnowPro certification program, candidates are required to first create an account at the Snowflake certification portal, register for the available certification, and accept Snowflake's certification terms and conditions during the registration process. Once the certification application is approved, then click Schedule Your Exam With Pearson VUE, which will take you to Pearson VUE to complete the registration process and schedule the exam.

Follow these steps to create an account with Snowflake at Pearson VUE:

1. Determine if you want to take your exam in person at a testing center or online. The online proctoring (OP) environment allows you to take the exam from the location of your choice such as your home or office. You can use your own computer for the exam in a locked-down browser. A proctor will monitor you via screen-sharing applications and your webcam (either an internal or external camera). Additionally, you must have the ability to install software for the online exam to launch.

2. Go to Pearson VUE and create your account. Figure 1-3 shows the login screen for a Pearson VUE trial account.

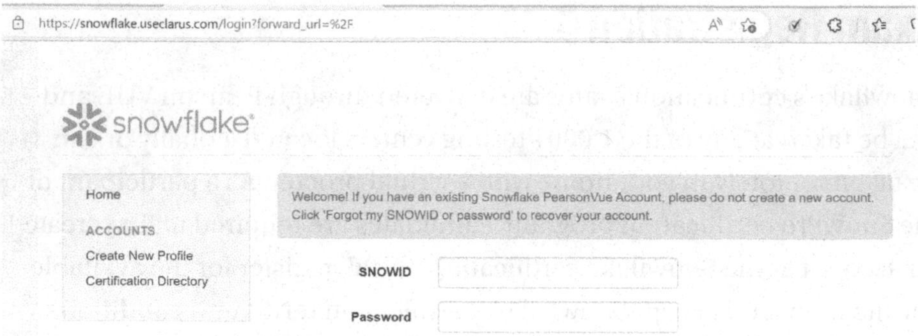

Figure 1-3. *Pearson VUE trial account*

1. Review and respond to Pearson VUE's privacy policy acceptance terms.

2. Create a web account. Time zone selection will be based on the address entered on your profile.

3. Schedule the exam by selecting the exam name under the pre-approved exams.

4. Locate the exam and check the delivery options.

5. Select the delivery option, date, and time you want to take the exam and agree to the Snowflake certification agreement. Here, the time you choose is the starting time for the exam. You should begin your check-in process no earlier than 30 minutes prior to the starting time.

6. Follow the payment instructions and click Submit.

Go to this link for any changes in the registration process: `https://learn.snowflake.com/courses/course-v1:snowflake+CERT-EXAM-REG+A/about`

Canceling or Rescheduling the Exam

If you are planning to take the exam at a testing center, then you must reschedule or cancel your registration at least 24 hours before the actual start time of your exam to avoid losing the exam fee. No shows, cancellations, and reschedules within 24 hours of your scheduled exam time will incur an additional fee or the forfeiture of your exam fee altogether. In the case of online proctoring, you can reschedule or cancel your registration any time before the start of the exam.

You are allowed to reschedule only up to four times before you are asked to cancel your exam.

Please note that these policies change, so always refer to the details in your registration email or the following link for the latest canceling or rescheduling policy from the vendor:

```
https://learn.snowflake.com/courses/course-
v1:snowflake+CERT-PRG-POL+A/about
```

Exam Retake Policy

If you fail your first attempt to pass a Snowflake certification exam, you must wait 7 calendar days from the date of your last attempt to retake the exam. In a 12-month period, you are allowed to retake a given exam four times. Each exam attempt will require payment of the exam fee.

If you have passed an exam and achieved certification, you will not be able to retake the same exam but can take the shorter recertification exam or another Advanced exam.

Understand the Exam Structure

Snowflake architects are in high demand with top organizations on a constant lookout for professionals who have deep knowledge of Snowflake. Given the growing popularity of Snowflake, being able to demonstrate your architect knowledge will be advantageous to your career.

Passing this certification is not an easy task if you have not prepared well; you must give extra time and effort to pass this exam.

Exam Format

The Snowflake Advanced Architect certification is a 115-minute exam with 65 questions (multiple select, multiple choice). A passing percentage is 750 with scaled scoring from 0 to 1,000.

This exam covers four domains: Account & Security, Snowflake Architecture, Data Engineering, and Performance Optimization.

Exam Domain Breakdown

Let's understand the domain breakdown and weighting ranges (Table 1-1). You can consider this table to be a comprehensive list of all the content that will be presented on the examination. These details are from the Snowflake vendor certification portal located here:

```
https://learn.snowflake.com/courses/course-v1:snowflake+
CERT-ARC-GUIDE+A/about?_ga=2.160190585.595359158.1673775264-
2017084655.1662272615
```

Table 1-1. *Exam Domain Breakdown*

Domain	Percentage
Accounts & Security	30%
Snowflake Architecture	25%
Data Engineering	20%
Performance Optimization	25%

Here is more information about the domains:

- *Account & Security domain*: This section primarily focuses on security architecture, which includes services that support user access and management, data access and compliance, encryption, and network security. This domain is divided into the following subsections:

  a. Design a Snowflake account and database strategy, based on business requirements.

  b. Design an architecture that meets data security, privacy, compliance, and governance requirements.

  c. Outline Snowflake security principles and identify use cases where they should be applied.

- *Snowflake Architecture domain*: This section is a quick win if you are clear on the Snowflake architecture, different layers, how caching results are stored, the size of micro partitions, data sharing and exchange methods, and SQL syntax for querying metadata. Some questions in this section are straightforward and a repeat of the SnowPro Core exam. Hence, a thorough review of what you studied for the SnowPro Core exam can help you here. This domain is divided into the following subsections:

a. Outline the benefits and limitations of various data models in a Snowflake environment.

b. Design data sharing solutions, based on different use cases.

c. Create architecture solutions that support development life cycles as well as workload requirements.

d. Given a scenario, outline how objects exist within the Snowflake object hierarchy and how the hierarchy impacts an architecture.

e. Determine the appropriate data recovery solution in Snowflake and how data can be restored.

- *Data Engineering domain*: As the name suggests, this section focus on the end-to-end aspects of data engineering. This domain is divided into the following subsections:

  a. *Kafka connector*: Focus on ingest flow for Kafka using the Kafka connector including its overall architecture, how many topics, internal stage, etc.

  b. *Data loading and unloading*: You should have a strong understanding of COPY INTO options, both for data loading and for unloading;

understand how Snowpipe is different from
bulk data loading; and know the best practices.
Prepare for simple topics like identifying an
option to unload one file or complex topics like
reading a real scenario and picking the best
possible option.

c. *Views*: Understand different types of views
(Regular/MV/Secure), know when to create
materialized views (MVs) versus regular
views, and be able to outline the benefits of
MVs, corresponding properties, downsides/
limitations, and use cases of MVs, regular views,
and external tables.

d. *Time Travel and cloning*: Define Time Travel
features, data replication, and failover and
understand how cloning works. Understand
access control privileges for cloned objects,
which objects that can be cloned, and how to
use different commands.

e. *Data pipeline in Snowflake*: This section
includes how to build an ETL data pipeline in
Snowflake using streams and tasks and different
components. It also covers how Snowpipe
works, the concept of change data capture,
and the SQL syntax used to create and clone a
stream and task.

11

- *Performance Optimization domain*: This domain focuses on performance tools, best practices, and appropriate scenarios in which they should be applied. This domain is divided into the following subsections:

  a. *Query profiles and tuning*: This is an important topic. You should understand how to read a query profile, identify bottlenecks, outline recommendations, and cut down query processing time in different scenarios.

  b. *Clustering*: You need to have a very good understanding of how micro-partitions work and how to interpret clustering metrics. This includes scenario-based questions (along with query profiles), clustering depth, strategies for selecting clustering keys, reclustering, and system functions to monitor clustering information for tables.

Scaled Scoring

Snowflake ensures you are evaluated fairly when you take the exam. Statistical analysis is used to set the passing scores, and scaled scoring models ensure consistency across multiple exam forms, item difficulty levels, and versions. The exam uses a scale of 0 to 1,000 with a passing scaled score of 750.

Exam Results

At the completion of the exam, candidates receive an emailed score report that contains important information regarding the outcome of the exam.

If you pass the exam, then your transcript will record the exam as passed. You will also receive an email from Snowflake's exam delivery vendor that contains your score. Within 72 hours of passing your exam, you will receive an email from Credly with a digital badge.

If you fail, then you can use the scoring feedback to focus on specific areas and retake the exam once you are ready.

Exam Tips

Based on my experience of taking various certification exams, here are some general tips that can help you prepare for the exam:

- *Make a study plan*: Once you are ready to prepare for an exam, it is important that you do not lose focus, so following the right study plan is extremely important.

- If you're new to Snowflake, then prepare a few months ahead. However, if you have been working in Snowflake for some time already and have a good grasp of the technology, then you might need less time.

- *Practice*: Since this is an advanced exam, it is imperative that you create a Snowflake trial account and do sufficient practice as it helps you prepare for hands-on questions.

- *Time management*: You need to manage your time so you can go through every question once and have some time for review toward the end too. On this exam there are 115 questions that you need to answer in 65 minutes, so you have approximately 1 minute and 70 seconds for each question.

When you start an exam, you sometimes get a feeling that you know nothing. This happens because of the excessive stress we are under while preparing for an exam. My suggestion is that if you are not clear about any answer, mark the question and move on to the next, returning to it later once you have answered all the questions you know.

It is advisable that you save multiple-select and scenario-based questions to the end since they need some time to process. Also, ensure you have sufficient time at the end of the exam to double-check your answers.

- *Attempt to answer every question*: Remember, there is no negative marking, so answer every question.

- *Multiple-select questions*: Start by eliminating clearly incorrect answers. Then, of the remaining options, identify the one that is clearly either first or last in the sequence. You have thus minimized the options you have to choose from and thereby minimized your risk of being incorrect.

- *Relax*: Sleep well before the exam, and when exam day finally arrives, relax. It is important to stay calm during the exam and go through every question carefully. Every certification exam has some passing criteria, so even if you are unable to answer a few questions, it is fine (as long as you are in a passing score of 750 with scaled scoring).

- *Read whitepapers and blogs*: To have a sufficient understanding of different case studies, you should go through the whitepapers and blogs available in the Snowflake resources.

Please refer to these additional free resources that can help you prepare for the exam:

https://quickstarts.snowflake.com/guide/getting_started_with_snowflake/index.html#0

https://learn.snowflake.com/courses/course-v1:snowflake+SPSG-ARA+A/about

https://community.snowflake.com/s/article/Caching-in-Snowflake-Data-Warehouse

www.snowflake.com/about/webinars/

https://community.snowflake.com/s/topiccatalog

Summary

In this chapter, we covered all the areas that will help you prepare for Snowflake Advanced Architect certification exam, which includes prerequisites for the exam, how to create a trial account, how to access resources, exam cost and validity, registration steps, exam format with a focus on domains and subsections, and a quick overview of why you should take this exam.

In the upcoming chapters, we will do a technical deep dive into various sections of the exam. Refer to the tips section at the end of every chapter for best practices and relevant links to Snowflake blogs.

CHAPTER 2

Snowflake Architecture and Overview

There are three service models for cloud computing: infrastructure as a service (IaaS), platform as a service (PaaS), and software as a service (SaaS). To start with, IaaS is where users access infrastructure (storage/warehouse) available on the cloud through a third-party provider (as a service). They can purchase the required infrastructure that they can scale up and down as needed. Similarly, PaaS is where a third-party provider provides an integrated solution that includes hardware and software on its infrastructure (as a service). It enables users to create their own apps without worrying about the environment and associated hardware/software updates. SaaS involves delivering an entire application (software) as a service without any need to install software on specific machines.

As a true SaaS platform, Snowflake combines the capabilities of a traditional data store and data lake with the benefits of the cloud. It offers dynamic computing power with consumption-based charging that can scale up for executing large queries and scale out for concurrency while providing exceptional performance.

© Ruchi Soni 2023

R. Soni, *Snowflake SnowPro™ Advanced Architect Certification Companion*, Certification Study Companion Series, https://doi.org/10.1007/978-1-4842-9262-4_2

Snowflake provides a plethora of services and features to support user data migration and modernization journeys. One of the main reasons for the popularity of Snowflake is its hybrid architecture.

In this chapter, we will do a deep dive into Snowflake shared-disk and shared-nothing architectures. The Snowflake architecture separates storage and compute, enabling users to use and pay for storage and computation independently, which means that users can scale up or down as needed and pay for only the resources they use. Snowflake uses massive parallel processing (MPP) compute clusters to process queries, and its data-sharing functionality supports real-time secure data sharing. The MPP technique refers to multiple processors (each with its own OS and memory) working on different sections of the same user program for faster performance. It involves a leader node that maintains metadata regarding compute nodes and communicates with them for the execution of different parts of the query. This is explained in Figure 2-1.

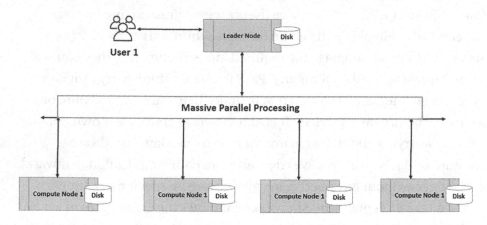

Figure 2-1. MPP

Snowflake Data Cloud

Snowflake is built from the ground up for the cloud. It runs completely on a cloud infrastructure and presently can be hosted on any MAG (Microsoft, Azure, Google) platform, which includes AWS, Azure, and GCP.

Snowflake provides unlimited real-time storage and computing along with the desired level of concurrency. As mentioned, it is a true SaaS platform where users don't have to worry about hardware installation, which makes it extremely useful for organizations and users with no dedicated resources. Users do not have to worry about software upgrades since Snowflake manages software installation and updates.

Snowflake enables users to run their workloads on one platform, which includes data sharing, data lake, data engineering, data science, and consumption. It also manages data and virtual warehouse upgrades. This means users can simply provision the server for an instance in the cloud without worrying about installation and configuration, and the system will be up and running in a few minutes. This reduces the time spent on installation and configuration, enabling users to rapidly scale their customer base. Snowflake combines a SQL query engine with a hybrid architecture. It separates storage and computing providing cost and speed benefits for users and supports the creation of an enterprise analytic database with additional capabilities. In this chapter, let's understand the Snowflake hybrid architecture.

Big Data Architecture Patterns

Let's first start with a discussion of two main architectural patterns: shared nothing and shared disk. In shared-nothing architecture, data is partitioned and spread across multiple nodes with each node managing the data it holds and not sharing it with any other node. Each node has a processor, memory, and disk. One node does not share memory or storage with another node, and communication among nodes is managed by the network layer. The main benefits of this approach are fault tolerance and scalability. The shared-nothing architecture works effectively in a high-volume and read-write environment. However, cost and performance are two main challenges to manage if users want to go with this option.

On the other hand, in shared-disk architecture, data is accessible from all nodes, allowing any node to read or write any portion of data. It is commonly used in distributed computing in which all nodes have their own private memory, but each node shares the same disk. Shared-disk systems are difficult to scale. Now, since all the nodes share the same disk, users should keep track of changes made (through a leader node) to ensure every node has a consistent view of data. This approach is useful for applications that are difficult to partition or require limited shared access to data. Figure 2-2 shows a diagrammatic view of these two different architectural approaches.

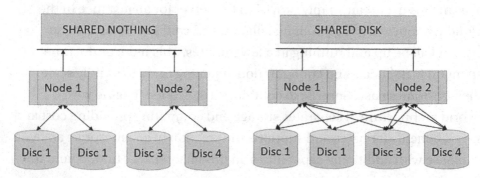

Figure 2-2. *Shared-nothing versus shared-disk architecture*

Typical data workloads can be OLTP and OLAP; OLTP is used for transactional data, and OLAP is used for analytical processing. A few examples of OLTP are credit card payment processing, reservation systems, etc. OLAP requires creating a data warehouse to store data and run user analytics to drive patterns, e.g., a customer sales trend report. Recently, Snowflake has announced a new feature called Unistore that combines both of these in a single platform that brings a unified approach to data governance and real-time analytical processing.

Snowflake Architecture

Now since we are well versed in the two main big data architecture approaches, this is the right time to talk about the Snowflake hybrid architecture (a combination of shared-nothing and shared-disk architecture) giving consumers the best of both. Just like shared-disk architecture, Snowflake uses a central data repository that can be accessed by all nodes, and like shared-nothing architecture, Snowflake uses massive parallel processing (MPP) virtual warehouses, with each node storing portion of the dataset locally, which gives performance and scale-out benefits. In simple words, MPP refers to using multiple compute instances to perform large-scale parallel computations.

Users can quickly scale out and scale up in Snowflake within seconds. Additionally, Snowflake comes with four different editions: Standard, Enterprise, Business Critical, and VPS. Each edition has a set of features (built on top of the previous edition), and users can choose a specific edition based on organizational needs. Snowflake also provides two user interfaces: Snowsight (a web platform) and SnowSQL (a command-line client). Snowsight is the latest web interface provided by Snowflake to replace the traditional SQL Worksheet and provides many additional functionalities such as automatic contextual statistics and data visualizations. SnowSQL is the command-line interface provided by Snowflake to execute SQL statements. It can execute in batch mode or run as an interactive shell.

Snowflake has a three-layer architecture as follows:

- *Storage layer*: This is the lowest layer where data is physically stored (cloud storage). Snowflake organizes data as compressed micro-partitions, which are adjoining units optimized for storage with each micro-partition containing between 50 MB and 500 MB of

21

uncompressed data. Data is organized in a columnar fashion, and Snowflake stores metadata about all rows stored in a micro-partition.

Snowflake takes care of creating and managing these micro-partitions along with other aspects of data storage. Since this layer is mapped to cloud storage, the cost of storing data in Snowflake is less and varies from approximately $23/TB per month in capacity storage to $40/TB per month for on-demand storage (with some change based on services and region).

- *Compute layer*: As the name suggests, this is the layer where queries are executed. Snowflake uses virtual warehouses to process queries, which are MPP compute clusters with compute nodes allocated by Snowflake. Consider the warehouse as an independent compute cluster with CPU, memory, and temporary storage, to perform the required operations. Snowflake provides amazing flexibility where warehouses can be started, stopped, or resized at any time to accommodate customers' computing needs based on the type of operations being performed, which really helps in query performance. There is also a provision for auto-suspend and auto-resume to limit idle time.

  Running a warehouse consumes credit, which increases as we increase warehouse size. Snowflake utilizes per-second billing (consumption-based). Snowflake also supports creating multicluster warehouses for allocating additional clusters to improve user performance/concurrence of queries.

- *Cloud services layer*: This layer takes care of all the services to coordinate across Snowflake (managing end-to-end workflow), which includes authentication, access control, metadata management, and query optimization and runs on a Snowflake-managed virtual warehouse.

Figure 2-3 shows these three layers and how they interact with each other.

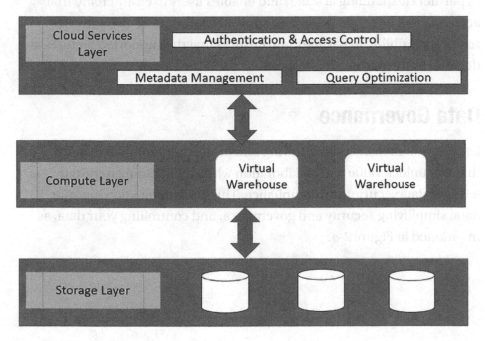

Figure 2-3. *Snowflake architecture*

Other Architecture Considerations

Snowflake delivers a single and seamless experience across clouds thereby eliminating data silos. It allows you to integrate data from a wide range of data sources and types (structured, semistructured, and unstructured) and use it to power use cases across domains. Many companies offer their data or data services delivered without data ever changing hands using the Snowflake Data Marketplace. Snowflake also has a big ecosystem of partners (expanding at scale) and enables users to easily create trial accounts and integrate them with Snowflake services. It also provides required flexibility to partners and ensures that the supported features are decided by the partners themselves.

Data Governance

Data governance includes knowing and protecting your data in a way that can unlock value and collaboration while maintaining required levels of data security and compliance. This is all about knowing your data, simplifying security and governance, and controlling your data, as mentioned in Figure 2-4.

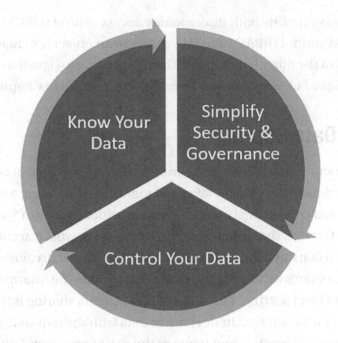

Figure 2-4. Governance in Snowflake

Snowflake provides a proper governance framework. In Snowflake, data can be stored in database/schema/tables in immutable (data cannot be deleted or updated) format with metadata. An information schema is rich in metadata and can be used for data classification, governance policies can be created once and applied across all datasets, sensitive data can be masked at the column level using native capabilities or external tokenization, row-level security can be used to restrict the results of a query, sensitive data can be tracked for effective data governance, external tables can be used to expose data for visibility outside Snowflake, there is native connectivity to data governance tools, etc. Here enterprise-wide data governance is achieved using the metadata, enabling business users to easily find, understand, and collaborate.

Snowflake supports both discretionary access control (DAC) and role-based access control (RBAC). In DAC, access to an object is managed by the owner. On the other hand, in RBAC, privileges are assigned to roles that are assigned to users. We will discuss these in detail in Chapter 14.

Secure Data Sharing

Secure data sharing allows object sharing between Snowflake accounts through the Snowflake services layer and metadata store using a network of provider and consumer accounts. A data provider refers to a Snowflake account that creates shares that are used by other Snowflake accounts. The accounts that consume these shares are called *consumer accounts*.

Consider shares as Snowflake objects that contain the information required for object sharing. The beauty of secure data sharing is that it enables quick access (zero latency) to live data with the required level of governance and controls and removes the risk of processes failing, etc. Here data is not copied to the target account, so there are no storage requirements at the consumer end for data that is shared by the producer. However, the consumer account will incur compute resource charges for querying the data. We will be discussing this concept in detail in subsequent chapters.

Tip   For data, sharing accounts should be in the same region and cloud. For cross-region/cloud data sharing, users should use database replication. I explain this feature in detail in later chapters.

Securable Object Hierarchy

A securable object refers to an entity with access. At the top of the hierarchy is the organization; objects such as tables, views, functions, and stages are contained in a schema object; and a schema object is contained in a database that is contained in an Account object that is contained in the organization. Each securable object has a single role associated with it. To create the object, that role can be assigned to users for object-shared control. This is explained in detail in Chapter 14.

Cloning

While working on projects, we need to copy all or part of production data to create dev and test environments. Typical use cases include testing part of code on a production-like environment or giving access to data scientists to play around with data. In a typical data store, creating a copy of production data will take time; we would need space to keep the copy of data, incurring data maintenance overhead to maintain two copies of data, etc. This has a major impact on the overall project cost.

Snowflake solves these challenges by cloning, also commonly called a *zero-copy clone*. Here users can clone their tables, schema, databases, and other objects without creating additional copies. Initially when users run a clone command, Snowflake creates a new entry in the metadata store for the new clone. However, once cloned, cloned objects are independent of the source, which means that changes in the source object or clone object are independent of each other. Also, a zero-copy clone does not take additional storage unless you make some changes to the clone. This is invisible to users and managed by the Snowflake services layer using metadata tables. We will be discussing cloning in detail in Chapter 8.

Tip   You cannot clone account-level objects including users, roles, grants, virtual warehouses, and resource monitors. Except for tables where we have the option to copy privileges to a clone for other objects, their clones do not automatically have the same privileges as their source objects, and we should explicitly grant them once a clone is created.

Micro-Partitions and Clustering

Snowflake organizes data as compressed micro-partitions, with one micro-partition containing groups of rows in a table organized in columnar format that is optimized for storage. These micro-partitions are created automatically by Snowflake using the ordering of data as it is inserted. Data is compressed within micro-partitions based on the compression algorithm determined internally by Snowflake, which also enables effective pruning of columns using micro-partition metadata. As data is inserted in tables, Snowflake keeps track of micro-partition clustering information in metadata tables and later uses this clustering information to avoid unnecessary scanning of micro-partitions during querying (query pruning), improving query performance.

Tip   Clustering depth is a way to determine if a table is partitioned. Now a table with no micro-partitions has a clustering depth of O. We discussed clustering depth and associated functions in detail in Chapter 7.

Caching

As we know, caching is a popular performance-tuning option that involves storing data in a temporary location for faster retrieval, resulting in resource optimization and performance tuning. Like the three-layer architecture in Snowflake, there are three layers of cache too. Each layer has its own significance and helps in overall performance tuning.

- *Result cache*: This is part of the cloud services layer and is available across virtual warehouses. If the underlying table data has not changed, then the results of a query

returned to one user are available for any other user
on the system who executes the same query since the
result cache is used.

- *Warehouse cache*: When users execute a new query
  and data is not available in the results cache, then data
  is retrieved from the remote disk storage and cached
  in the SSD and memory of the virtual warehouse. This
  data remains active until the warehouse is active. When
  users execute a query and data is already available in
  the warehouse cache, then there is no need to access
  the storage layer.

- *Remote disk cache*: Although it is called the *remote disk
  cache*, it actually refers to long-term storage (cloud
  storage) where data is physically stored.

Tip   There is another cache called a *metadata cache* that contains
metadata about tables and micro-partitions.

Figure 2-5 explains the three layers and how they
communicate with each other.

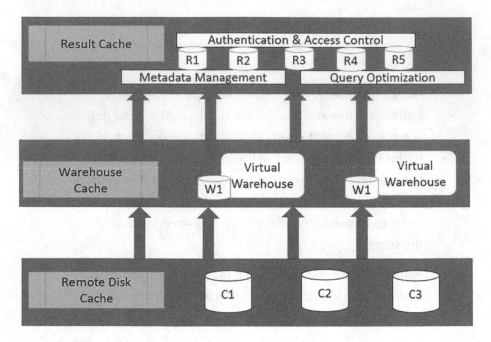

Figure 2-5. *Caching in Snowflake*

Tip   When a user executes a query, Snowflake internally validates if the results are available in the result cache. If not, it checks the warehouse cache and eventually the remote disk cache.

Summary

In this chapter, we discussed the three-layer Snowflake architecture including other architecture considerations such as cloning, secure data sharing, micro-partitions, and caching at a high level. We also discussed important commands/functions and parameters. In the next few chapters, we will get into a few of these features in detail and explain additional features that are important for the Advanced Architect exam.

CHAPTER 3

Kafka Connectors and Client Drivers

Snowflake provides multiple options to connect with its ecosystem. This includes a list of partners and technologies certified to provide native connectivity to Snowflake; Snowflake-provided clients to create trial accounts with identified partners; and Snowflake-provided drivers and connectors for Python, Spark, JDBC, ODBC, etc.

In this chapter, we will discuss the Snowflake Kafka connector, which is used to read data from Kafka topics in detail, and we will give an overview of a few other client drivers.

Snowflake Kafka Connector

Apache Kafka is a distributed event streaming platform and uses a pub-sub (publish and subscribe) model to process streams of records for real-time streaming data pipelines and mission-critical applications. It is an open-source system in Java and Scala developed by the Apache Software Foundation.

In the traditional Kafka, the pub-sub model source publishes messages to a topic, and targets subscribe to these topics to access messages. The relationship here is many-to-many, which means an application can publish to multiple topics, and vice versa.

The Snowflake Kafka connector reads data from Kafka topics and writes into target Snowflake tables. Assuming each Kafka message as a record, the Kafka topic provides a stream of records that are inserted in tables. The Snowflake Kafka Connector supports both the Confluent package version (for the Confluent platform) and the open-source version (OSS). However, in both cases, the basic principles for access remain the same.

Structure of Table Loaded by the Kafka Connector

In general, a Snowflake table loaded by the Kafka connector contains two Variant columns.

1.   RECORD_CONTENT with Kafka message

2.   RECORD_METADATA with metadata associated with the message

If Snowflake creates tables for the Kafka connector, then it contains only these two columns, but if a user creates a table, then they can add more columns; however, all the additional columns must support NULL values.

Each Kafka message is passed to Snowflake in JSON/AVRO format. This message is stored in the RECORD_CONTENT field.

The RECORD_METADATA column contains the following metadata for each Kafka message:

1.   *Topic*: Kafka topic name

2.   *Partition*: Number of the partition within the topic

3.   *Offset*: Partition offset

4.   *CreateTime/LogAppendTime*: Timestamp of the message in the Kafka topic

5. *Key*: The key for Kafka KeyedMessage

6. *Schema_id*: Schema's ID in the registry (valid only in the case of using Avro with a schema registry)

7. *Headers*: User-defined key-value pair

Kafka Connector Workflow

Let's understand the Kafka connector workflow that explains how the Kafka connector is used to move data from the publisher to Snowflake tables.

Here are the steps involved:

1. The Kafka configuration file contains parameters such as the application name, list of topics, Snowflake login credentials, user authentication details, Snowflake database name, schema details, etc.

Tip   The Kafka connector can read messages from multiple topics, but the related tables should be part of one database/schema. Additionally, since the configuration file contains security-related information, users should store the file in a secure location and set the correct privileges to limit file access.

---

2. The Kafka connector uses information stored in the configuration file to subscribe to one or more Kafka topics.

3. For each topic, the Kafka connector creates the following objects:

- One internal stage to store data.

- One pipe for each topic partition to ingest data.

- One table for each topic. By default, Snowflake assumes the same name for the table and the topic. To ensure both are different, users should use the topic2table.map parameter in the Kafka configuration file.

Tip   If the specified table does not exist, the Kafka connector creates a new table with only two columns: RECORD_CONTENT and RECORD_METADATA. For an existing table, the connector adds the RECORD_CONTENT and RECORD_METADATA columns (other columns should be nullable or else it throws an error).

The Kafka connector triggers Snowpipe to ingest data from the internal stage. It copies a pointer that refers to the data file into a queue and uses a virtual warehouse to load data from the internal stage to the target table.

Once data is loaded into the Snowflake tables, the Kafka connector removes the file from the internal stage. If there are any failures during data loading, the Kafka connector throws an error and moves the related data file into the table stage. Snowflake polls the insertReport API for one hour. If the status of an ingested file fails within this hour, the files are moved to a table stage. This API retains 10,000 of the most recent events for a maximum duration of 10 minutes only.

Figure 3-1 gives a pictorial representation of the previously explained process.

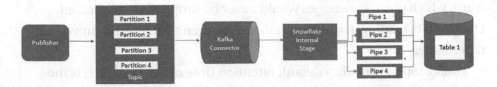

Figure 3-1. *Kafka Connector workflow*

While using the Kafka connector, the costs involved are related to Snowpipe processing time and data storage. However, as we know, the data storage costs in Snowflake are minimal, so most of the cost of using the Kafka connector comes from the Snowpipe processing time.

Shut Down the Kafka Connector

With the right privileges, users can shut down the Kafka connector if it is no longer in use and drop the associated Snowflake objects. This includes the following:

- Drop stage

- Drop pipes

- Drop tables

Users should be careful while deciding the next steps for the internal stage. Each internal stage stores data files and the state information used for data movement from Kafka to the table. If both are preserved and the Kafka connector is restarted, it resumes from the point where it was left; otherwise, the Kafka connector would resume from the beginning.

Managing Fault Tolerance

As the name suggests, *fault tolerance* is the ability of the system to continue its normal functioning in the case of any failure. The Kafka connector is fully fault tolerant and ensures that there is a single version of the truth, which means messages would never be dropped or duplicated. During data loading, if the error is detected, then the record is moved to a table stage.

Kafka topics provide a default retention time of 7 days, which is the time for which they would retain and load records (if the system is idle). Similarly, there is a limit to the storage space, and if this limit is exceeded, then messages would be lost. Both of these are configurable properties.

Recommendations While Configuring Kafka Connectors

The following are a few recommendations given by Snowflake to consider while configuring Kafka connectors:

- For each Kafka instance, create an individual user and role and assign it as a default user role for access privileges to be revoked individually.

- To ensure Kafka connector compatibility, Snowflake recommends using a Kafka Connect API version between 2.0.0 and 2.8.1.

- While configuring Kafka Connect Cluster, ensure that the cluster node contains enough RAM (the minimum recommended amount is 5 MB per Kafka partition).

- For better throughput, keep the Kafka connector and user Snowflake account in the same cloud provider region.

Other Client Drivers

Snowflake provides additional drivers as an interface to connect with related applications and perform associated operations to connect with Spark, Kafka, Python, JDBC, ODBC, Node.js, .NET, etc.

Snowflake Connector for Python

Snowflake Connector for Python supports versions 3.6, 3.7, 3.8, and 3.9. It is installed using a standard Python package installer (pip) and uses the HTTPS protocol to establish connectivity. It automatically converts from Snowflake to native Python data types.

Snowflake Connector for Spark

The Snowflake Connector for Spark supports Spark 3.0, 3.1, and 3.2 versions. There is a separate version of the Snowflake connector for each version of Spark, and it is used for moving data between the Snowflake cluster and the Spark cluster. There are two data transfer modes provided by the Snowflake connector for Spark.

- *Internal transfer*: Here the storage location is managed automatically by Snowflake for data movement between the two systems.

- *External transfer*: Here the storage location to facilitate data movement between the two systems is created and managed by the user during connector installation. Additionally, the user should manually delete data files created during data transfer.

Snowpark

Snowflake has recently launched Snowpark, which allows teams to collaborate on the same single copy of data, while natively supporting everyone's programming language of choice. It allows users to perform work within Snowflake rather than in a separate Spark compute cluster without the need to move data to another system. It also supports the pushdown of all operations, including Snowflake UDFs.

Figure 3-2 explains how you can build a data pipeline for data scientists using Snowpark.

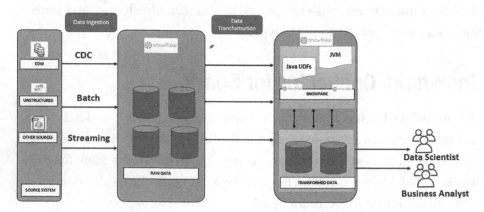

Figure 3-2. *Data pipeline using Snowpark*

Snowpark enables using Python, Java, or Scala with a familiar data frame and custom function support to build efficient pipelines while working inside Snowflake's data cloud. You get the benefits of Snowflake governance, security, and performance at scale.

Summary

The first step when creating a data pipeline is to connect with the source systems. In this chapter, we discussed a few important connectors and drivers that form the extended ecosystem for connecting to Snowflake. We discussed in detail the Snowflake connector for Kafka since that is extremely important to understand from an exam perspective. In the next chapter, we will focus on how we load data in Snowflake tables.

CHAPTER 4

Loading Data into Snowflake

In previous chapters, we discussed various options provided by Snowflake to connect with its ecosystem. Now, assuming a connection with the source system has been established, let's dive into various options to load data into Snowflake.

Typically, data can be loaded in batches, in micro-batches, and in real time. Batch processing can be used in cases where businesses need to load large amounts of data without any need for real-time data availability, e.g., to create a weekly report of the payroll function. A micro-batch is used to load data in small batches that are processed and loaded at regular intervals. They can be used for cases where a business needs reports that are refreshed at frequent intervals like every few minutes or hours. Real-time processing is used for cases where a business wants to create reports on data closer to the time it was created (almost in real time).

Depending on the volume and frequency of data, Snowflake provides two options for data loading: bulk versus continuous. As the name suggests, bulk loading is used to load bulk data (in batches) from a Snowflake internal/external stage to Snowflake tables, whereas continuous loading is used to load small volumes of data and incrementally make them available for analysis.

© Ruchi Soni 2023
R. Soni, *Snowflake SnowPro™ Advanced Architect Certification Companion*,
Certification Study Companion Series, https://doi.org/10.1007/978-1-4842-9262-4_4

Data Loading Types

The following are data loading types.

Bulk Data Loading

Bulk data loading is the most frequently used method to load data in Snowflake tables. It uses the COPY command for data loading from the internal/external stage (explained later) along with a virtual warehouse specified by the user. These warehouses are managed by users as per the required workloads.

Snowflake also supports these basic transformations while data loading using the COPY command:

- Reordering of columns

- Omitting columns

- Type casting

- Data truncation

Understanding Stage

In simple terms, *stage* refers to the location of data files in cloud storage from where data can be loaded into Snowflake tables. There are two types of stages in Snowflake.

- *External stage*: This refers to external cloud provider storage services that include Amazon S3, Google Cloud Storage, and Microsoft Azure. Consider an external stage as a database object with certain properties and access settings. Snowflake supports data loading from any of these cloud storage locations irrespective of the cloud platform of the Snowflake user.

Tip   A named external stage is a database object that stores the URL to files in the cloud storage along with information like settings to access the cloud storage account, format of staged files, etc.

- *Internal stages*: This refers to local storage maintained by Snowflake and can be used if users do not have access to an external cloud provider. Snowflake supports three types of internal stages.

  - *User stage*: By default, a user stage is assigned to each user to store and manage files. These files can be loaded into multiple Snowflake tables, so this stage should be used when data files are accessed by only one user but loaded into many tables. They are referenced using @~. For example, `list @~employee` can be used to list files in the `employee` user stage.

  - *Table stage*: As the name suggests, the table stage is available for each table object in Snowflake. It is created to store files managed by one or multiple users but is loaded in only a single table. This stage should be used for the quick loading of files into one table. They are referenced using @%. For example, `list @%customer` can be used to refer to table stage `customer`.

  - *Named stage*: The named stage is used to store files managed by one or many users and loaded into one or more tables. It is a database object created within a schema and follows the same security policies as other objects. Named stages should

43

be explicitly created using the CREATE STAGE command (not automatically created like user and table stages). Since they enable multiple users to access the same stage, they should be used for regular data loading involving multiple users and tables. They are referenced using @. For example, list @sample_stage can be used to view a list of files in sample_stage.

The following command is provided by Snowflake to create an external/internal stage:

```
CREATE OR REPLACE STAGE <stage_name>
<stage_params>

[ FILE_FORMAT = ( { FORMAT_NAME = '<file_
format_name>' |

TYPE = <file_format_type> <format_type_
options>]

[ COPY_OPTIONS = ( copy_options ) ]

Here
```

- Stage_params: This includes options for storage location (for external stage), encryption type, etc.

- file_format_type: This refers to the format of file, and the acceptable values here are CSV, JSON, AVRO, ORC, PARQUET, and XML.

- format_type_options: This includes options such as compression algorithm, record delimiter, field delimiter, etc.

- copy_options: Here users specify the next steps in case of error (continue/skip file), purge, etc.

Tip   As a best practice, staged files should be deleted from the Snowflake stage (REMOVE command). It reduces the number of files to be scanned by the COPY_INTO command to verify if existing files are loaded successfully from the stage thereby improving performance. Also, all files stored in Snowflake internal stages are automatically encrypted using AES-256 strong encryption.

Figure 4-1 gives a sequence of steps used to create the external stage SAMPLE_STAGE in SAMPLE_DB from the sample dataset present in an S3 bucket associated to my AWS account.

```
CREATE OR REPLACE DATABASE SAMPLE_DB;
CREATE OR REPLACE SCHEMA SAMPLE_DB.EXTERNAL_STAGE;
CREATE OR REPLACE STAGE SAMPLE_DB.EXTERNAL_STAGE.SAMPLE_STAGE
url='s3://snowflakebucket05/sample_data_1.csv'
credentials=(aws_key_id='XXXX' aws_secret_key='XXXX');
DESC STAGE SAMPLE_DB.EXTERNAL_STAGE.SAMPLE_STAGE;
```

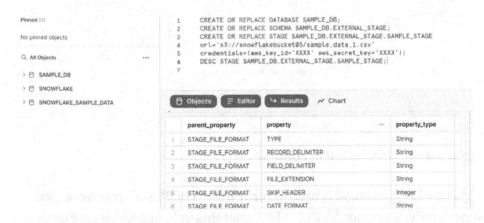

Figure 4-1. *Creating an external stage*

Users can create named file formats database objects to encapsulate format information for the data files.

It is a good practice to use named file formats in cases where users load the same type of data (same format). They can be used as input in all the places where users specify individual file format options.

Tip   If the file format is specified in multiple locations, the order of precedence used for data loading is first the `Copy Into Table` statement, then the stage definition, and finally the table definition.

Another important thing to understand is the difference between the `DESC STAGE` and `LIST` commands. Figure 4-2 explains the use of the `DESC STAGE` command for our `SAMPLE_STAGE`. It explains the property name, property type, and default value for stage properties.

```
DESC STAGE SAMPLE_DB.EXTERNAL_STAGE.SAMPLE_STAGE;
```

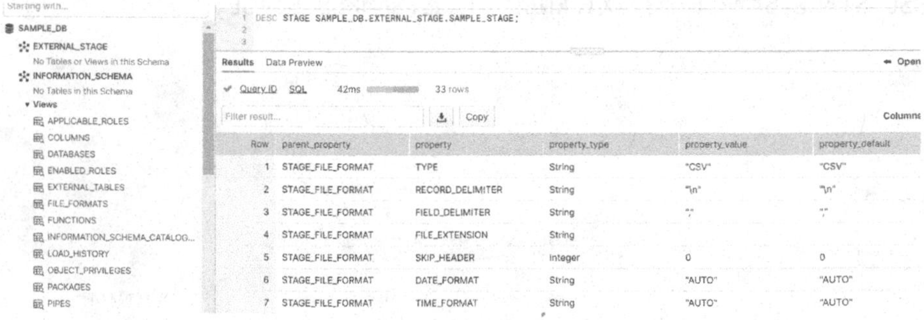

Figure 4-2. *DESC STAGE command*

The `LIST` command is used to verify the list of files users have staged successfully. Since our `SAMPLE_STAGE` contains only one file, if we execute the `LIST STAGE` command in Figure 4-3, we get the following output:

```
LIST @SAMPLE_DB.EXTERNAL_STAGE.SAMPLE_STAGE;
```

Figure 4-3. *LIST STAGE command*

Tip   Since an external stage refers to an external cloud provider
storage service, users should give the URL of the storage location
while creating an external stage. If a URL is not specified, then
Snowflake does not throw any error, but it creates an internal stage
by default.

Continuous Data Loading

This data loading option is designed for small volumes of data that should
be loaded quickly in Snowflake to ensure the latest data is available for
quick user analysis. Continuous data loading is done using Snowflake's
popular Snowpipe (a serverless computing model).

Once files are added to the stage, Snowpipe loads data within minutes,
which ensures users always have access to the latest data for analysis if the
AUTO_INGEST property of the pipe is set to TRUE (this is discussed in detail
in Chapter 12). Snowpipe uses compute resources provided by Snowflake,
which are automatically resized and scaled up or down as required
(serverless computing). This means users can run their queries and load
any file size with Snowflake managing computing and scaling up and
down based on Snowpipe load.

Tip Because of the serverless computing model of Snowpipe, accounts are charged based on their actual virtual warehouse usage. Snowflake tracks resource consumption with per-second/per-CPU core and converts this utilization into credits.

Data Loading Using the COPY Command

Users can use the PUT command to load files from their local system to the Snowflake stage (on the cloud). Once loaded, the COPY command enables loading data from the internal/external stage to Snowflake tables. It also supports transforming data while loading it into tables. This command gives multiple options, and understanding these options is very important for the exam. The following is the typical syntax of the COPY command:

```
COPY INTO <table_name>
FROM { stage }
[FILES=('f1','f2','f3',...)]
[ PATTERN = '<regex_pattern>' ]
[ FILE_FORMAT = (
{ FORMAT_NAME = '[<namespace>.]<file_format_name>' |
TYPE = { CSV | JSON | AVRO | ORC | PARQUET | XML } [
formatTypeOptions ] } ) ]
[ copyOptions ]
[ VALIDATION_MODE = RETURN_<n>_ROWS | RETURN_ERRORS | RETURN_
ALL_ERRORS ]
```

To demonstrate the use of the COPY command, Figure 4-4 shows the sequence of steps used to create the EMPLOYEE table and then use the COPY INTO command to load data from SAMPLE_STAGE created earlier into the mentioned table.

```
Create or replace table employee
(First_Name string,
Last_Name string,
City string,
County string,
State string,
ZIP string);
COPY INTO EMPLOYEE FROM @SAMPLE_DB.EXTERNAL_STAGE.SAMPLE_STAGE
file_format= (type = csv field_delimiter=',' skip_header=1)
```

Figure 4-4. *Using COPY INTO to load data into the table*

Understanding all the `COPY INTO` parameters (mandatory and optional) is important for exam as you should expect a few scenario-based questions on the topic. Let's discuss a few important options.

a) The following option is used by Snowflake to understand the compression algorithm of data. To load BROTLI-compressed files, users should explicitly use BROTLI compression instead of AUTO.

```
COMPRESSION = AUTO | GZIP | BZ2 | BROTLI | ZSTD |
DEFLATE | RAW_DEFLATE | NONE
```

b) The following specifies a list of data files for data
loading. These files should be available in the
staging area; otherwise, Snowflake follows the
option specified in the ON_ERROR parameter (the
default behavior is ABORT). There is an upper limit
on the number of file names that can be specified
here (1,000 files).

```
FILES=('f1','f2','f3',...)
```

c) This option refers to the type of file for data loading:

```
TYPE = CSV | JSON | AVRO | ORC | PARQUET | XML [ ... ]
```

d) The following option is used for validating data files
only (no data loading) and returns results based
on the validation option specified. Users can then
correct these files to ensure smooth data loading.

```
VALIDATION_MODE = RETURN_n_ROWS | RETURN_ERRORS |
RETURN_ALL_ERRORS
```

The following are the details:

- RETURN_n_ROWS: Here n refers to the number of rows to
  validate. This option validates these rows and fails at
  the first error encountered.

- RETURN_ERRORS: This returns all errors across all the
  files in the COPY command.

- RETURN_ALL_ERRORS: This returns all errors across all
  the files in the COPY command, including files that were
  partially loaded during an earlier load.

To demonstrate VALIDATION_MODE, Figure 4-5 explains what happens if we try to load data from a file in a table having fewer columns than the file.

```
create or replace table employee_error
(First_Name string,
Last_Name string,
County string,
State string,
ZIP string);
COPY INTO EMPLOYEE_ERROR FROM @SAMPLE_DB.EXTERNAL_STAGE.
SAMPLE_STAGE
file_format= (type = csv field_delimiter=',' skip_header=1)
VALIDATION_MODE = RETURN_ERRORS;
```

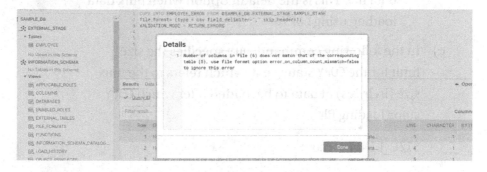

Figure 4-5. *Demonstrating VALIDATION_MODE*

Tip   To retrieve data loading history using the COPY INTO command, use the LOAD_HISTORY information schema view.

---

a)   In the following, <n> refers to the number of lines to skip at the beginning of the file. The default value is 0.

```
SKIP_HEADER = <n>
```

b) This parameter gives the error handling options for
data loading:

`ON_ERROR = CONTINUE | SKIP_FILE | ABORT_STATEMENT`

This includes the following:

- `CONTINUE`: This continues to load the file even after
the error.

- `SKIP_FILE`: If an error is encountered in a file, then
it is skipped from the data loading process. This is
the default option when using Snowpipe.

- `ABORT_STATEMENT`: The immediately aborts the
load operation as soon as the error is found in any
data file. This is the default option when bulk data
loading using `COPY`.

c) In the following command, `num` specifies the size
limit for the `COPY` statement, which refers to the max
size (in bytes) of data to be loaded after which `COPY`
stops loading files.

`SIZE_LIMIT= <num>`

d) This refers to the purging policy of data files in the
staging area after a successful data load. If set to
`TRUE`, then files are purged from the staging area
once the load operation completes successfully. If
the purge operation fails, no error is returned.

`PURGE= TRUE|FALSE`

e) The following is a Boolean operation that if set to
`TRUE` forces the `COPY` command to load all files,
regardless of whether they have been loaded

previously. This can result in data duplication
However, note that regardless of the value of this
parameter, Snowflake by default will not load the
same file within a 64-day window.

```
FORCE = TRUE | FALSE
```

Tip   To reload the staged file, set FORCE = TRUE or modify the file contents and stage it again.

---

f)   The following parameter is used to truncate text
that exceeds the column length during data load.
If the parameter value is TRUE, then the strings are
truncated; otherwise, it produces an error if the source
data value is more than the length of the target field.

```
TRUNCATECOLUMNS = TRUE | FALSE
```

Tip   TRUNCATECOLUMNS and ENFORCE_LENGTH provide the same functionality but with opposite behavior (TRUNCATECOLUMNS truncates for TRUE, and ENFORCE_LENGTH truncates for FALSE).

Snowpipe

Snowpipe is a data ingestion service for continuous data loading that loads data from files in micro-batches after they are added to a staging area (near real-time ingestion). In Snowflake terminology, a *pipe* is an object containing a COPY statement that identifies the source location of the data files and a target table. Being a serverless service for data loading, by using Snowpipe, users are billed based on the computing resources used by the service

Tip Pipes do not support the PURGE copy option, which means Snowpipe cannot delete staged files automatically when the data is successfully loaded into tables. To remove staged files, users should periodically execute the REMOVE command to delete the files.

Snowpipe Stage File Availability Options

Snowpipe supports two options to detect the availability of files in the stage.

- *Automate Snowpipe using cloud messaging*: This option uses a cloud event notifications service to notify Snowpipe when new data files are available in the staging area. Notifications identify the cloud storage event and include a list of filenames that are copied into a queue and loaded into the target table using Snowpipe. Kindly note that cross-cloud support is available for accounts hosted on AWS.

Tip Event notifications received while a pipe is paused are retained for 14 days, when Snowflake schedules it to be dropped from the internal metadata. If the pipe is later resumed, Snowpipe may process notifications older than 14 days on a best-effort basis.

- *Snowpipe REST API calls*: This involves calling a public REST endpoint with a pipe name and list of files for data loading. When new data files matching the list of files are available in the stage, they are queued for loading. It requires key-pair authentication with JSON Web Token (JWT) signed using a public/private key pair with RSA encryption.

Tip   Snowpipe tries to load files in FIFO order (older to latest), although there is no guarantee that files are loaded in the order in which they are staged.

Data Load Using Snowpipe

Snowpipe requires the Java/Python SDK. It supports both the named internal or external stage and the table stage.

Figure 4-6 explains how you create a pipe called EMPLOYEE_PIPE.

```
CREATE OR REPLACE SCHEMA SAMPLE_DB.PIPES;
CREATE OR REPLACE pipe SAMPLE_DB.PIPES.EMPLOYEE_PIPE
auto_ingest = TRUE
AS
COPY INTO SAMPLE_DB.EXTERNAL_STAGE.EMPLOYEE FROM @SAMPLE_
DB.EXTERNAL_STAGE.SAMPLE_STAGE
file_format= (type = csv field_delimiter=',' skip_header=1);
DESC pipe EMPLOYEE_PIPE;
```

Figure 4-6. *Create a pipe EMPLOYEE_PIPE*

Key-based authentication (public-private key pair) should be used for making calls to Snowpipe REST endpoints. Once data files are staged, the `insertFiles` REST endpoint should be used for data loading.

Tip Snowpipe uses file loading metadata, which contains the name and path of the loaded file associated with each pipe object to avoid data duplication.

Data loading using Snowpipe REST endpoints is a three-step process.

1. Data files are copied to a Snowflake internal/external stage.

2. Call the `insertFiles` endpoint with the pipe name and file list for data loading. When data files matching the list are available in the staging area, then they are moved to the data ingestion queue.

3. The virtual warehouse managed by Snowflake (serverless compute) loads data files from the ingestion queue to the target table.

Snowpipe Rest APIs

This is another important topic for the exam. The following are REST endpoints provided by the Snowpipe API:

- insertFiles: This REST API informs Snowflake about the files to be ingested. A positive response here means that Snowflake has confirmed the list of files for data loading. The POST can contain a maximum of 5,000 files, and each file path must be less than or equal to 1,024 bytes long (UTF-8). Here are the response codes to remember:

  - 200: Success, which means files are added to the queue for data ingestion

  - 400: Invalid format/limit exceeded failure

  - 404: Pipe name not recognized failure

  - 429: Request rate limit exceeded failure

  - 500: Internal error failure

Tip   The insertFiles response payload includes requestId, status elements, and errors (in case of failure) in JSON format.

- insertReport: This retrieves a report of files recently ingested in Snowflake. However, events are retained for a maximum of 10 minutes, and only the 10,000 most recent events are retained. Here are the response codes to remember:

  - 200: Success. This returns the last report.

  - 400: Invalid file format/limit exceeded failure.

- 404: Pipe name not recognized failure.

- 429: Request rate limit exceeded failure.

- 500: Internal error occurred failure.

- `loadHistoryScan`: This creates a report about files recently ingested into the table and the view history between two points in time. It can return a maximum of 10,000 items, but the user can issue multiple calls to cover the time range. Here are the response codes:

  - 200: Success, which returns load history scan results

  - 400: Invalid format/limit exceeded failure

  - 404: Pipe name not recognized failure

  - 429: Request rate limit exceeded failure

  - 500: Internal error occurred failure

Tip   The `loadHistoryScan` response payload of a successful response contains information about files recently added in Snowflake.

Summary

Data loading is an important part of any data pipeline. We discussed various options for loading data in Snowflake. This includes continuous and bulk data loading including a deep dive into Snowpipe and various parameters of the COPY INTO command. In the next chapter, I will explain how we can unload data from Snowflake.

CHAPTER 5

Data Unloading from Snowflake

In previous chapters, we discussed different options provided by Snowflake for loading data into our ecosystem. Like with data loading, Snowflake provides a data unloading option that includes the bulk export of data from the database into files. In this chapter, we will discuss the details of this data unloading process.

Bulk Unloading

Bulk data unloading from Snowflake is a two-step process, as mentioned here:

1. Just like data loading, use the COPY INTO command to copy/unload data from Snowflake tables into the Snowflake stage.

2. Move files from the Snowflake stage into your local system. This can be done using the GET command for the internal stage and interfaces provided by the cloud provider for the external stage.

© Ruchi Soni 2023
R. Soni, *Snowflake SnowPro™ Advanced Architect Certification Companion,*
Certification Study Companion Series, https://doi.org/10.1007/978-1-4842-9262-4_5

Figure 5-1 explains how you can use a combination of the COPY INTO and GET commands to unload data from Snowflake. Here, the COPY command is used to unload rows from Snowflake tables into the Snowflake stage (internal/external), and the GET command is used to download these files from the stage to the user's local machine.

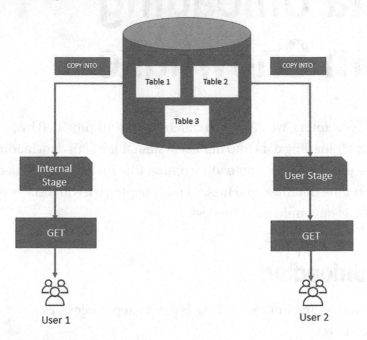

Figure 5-1. *Data unloading from Snowflake*

For data security, Snowflake automatically encrypts data (using 128-bit keys) while unloading data to a staging area and decrypts it when downloading to the user's local system. For stronger encryption, users can enable 256-bit keys with additional configuration. Additionally, data files unloaded to specified cloud storage can also be encrypted using a user-provided security key. Irrespective of the file format, UTF-8 encoding is used for unloaded files.

Data Unloading Using the COPY Command

The COPY command enables unloading data from Snowflake tables into the internal/external stage to Snowflake tables. We already discussed the COPY command syntax and parameters (mandatory and optional) in the previous chapter. The beauty of Snowflake is that it allows users to use the full range of SQL functionality within the SELECT statement within the COPY INTO statement including JOIN conditions, which can be used to combine data from multiple tables during data unloading.

Let's understand data unloading using one example. Assume we have an employee table and want to move the contents of that table to an external stage called SAMPLE_STAGE, as shown in Figure 5-2.

```
COPY INTO @SAMPLE_DB.EXTERNAL_STAGE.SAMPLE_STAGE
FROM
(SELECT * FROM EMPLOYEE)
```

Figure 5-2. *Unloading data from table to external stage*

Hence, in this chapter, let's discuss the COPY command options used for data unloading.

- FILE_FORMAT = (TYPE = Delimited, JSON, Parquet...): This option specifies the format of the unloaded data file. Here values can be Delimited

for structured data and JSON, Parquet, etc., for
semistructured data. When unloading JSON files,
Snowflake outputs to the Newline Delimited JSON
standard format. JSON should be specified when users
are unloading data from the VARIANT column in tables.
Individual file format options can be specified in any of
the following places:

- In a table definition

- In a names stage definition

- Directly in the COPY command

- COMPRESSION = AUTO | GZIP | BZ2 | BROTLI |
ZSTD | DEFLATE | RAW_DEFLATE | NONE: This option
is used by Snowflake to understand the compression
algorithm to compress unloaded data files. The
default compression algorithm used is gzip, unless any
compression method is specified or the user disables
compression. For BROTLI-compressed files, users
should explicitly use BROTLI compression instead
of AUTO.

- NULL_IF = ('S1' ,'S2'...): This option can be used
for string conversion from SQL NULL to the first value in
the mentioned list.

- EMPTY_FIELD_AS_NULL = TRUE | FALSE: This is an
interesting option provided by Snowflake to distinguish
empty strings with NULL values and unload data with
empty strings from a table. When set to FALSE, this
option can be used to unload empty strings without
enclosing quotes.

- OVERWRITE= TRUE|FALSE: When set to FALSE, this
  option avoids overwriting files with the same name.
  If the value is TRUE, then it overwrites the file with the
  same name. There is no impact on the files that do
  not match the names of files mentioned in the COPY
  command.

- SINGLE= TRUE|FALSE: To maximize parallel processing,
  data is always unloaded into multiple output files. By
  default, the size of each output file is 16 MB, but users
  can change the output file size using the MAX_FILE_
  SIZE option (the maximum size of a file can be 5 GB).
  However, if users want to output data in a single file,
  then they should set SINGLE=TRUE.

- To ensure each filename is unique across parallel
  execution threads, users can specify a filename prefix
  that is assigned to all the files. If the prefix is not
  specified, then Snowflake prefixes filenames with
  <data_>. It also appends a suffix to the filename.

- MAX_FILE_SIZE = <n>: Users can change the output file
  size using this option. Here <n> stands for the new file
  size. As mentioned earlier, the default file size of each
  output file is 16MB, but users can change the output
  file size using the MAX_FILE_SIZE option (the maximum
  size of a file can be 5 GB).

- PARTITION BY <expression>: Here <expression> is
  used to partition data while unloading. Partitioning
  unloaded data into a directory structure in cloud
  storage can increase the efficiency for third-party

tool consumption. Users specify an expression where
the unload operation partitions table rows into
separate files.

- FIELD_OPTIONALLY_ENCLOSED_BY = '<character>' |
  NONE: This option can be used to enclose strings in the
  <character> (single quotes or double quotes) or NONE.

Tip   Instead of a table, Snowflake supports SELECT statements in
the COPY command.

Summary

You can consider this chapter as an extension of the previous chapter with
a focus on data unloading options. Snowflake supports the bulk export of
data into flat files or tables for data unloading. Just like with bulk import, it
uses the COPY INTO command for bulk export. In this chapter, we focused
on the key steps involved in data unloading and the important parameters
associated with the COPY INTO command specific to data unloading. From
an exam perspective, it is important to understand various parameters
available for data loading/unloading, best practices, and which is the right
option to choose based on the requested business scenario.

CHAPTER 6

Tables and Views

To store data in Snowflake, users first create a database that contains one or more schemas that includes various objects such as tables, views, etc. This Snowflake object hierarchy is discussed in detail in Chapter 14. Understanding different types of Snowflake tables and views and their specific business use case is very important for Advanced Architect exams. Hence, in this chapter, let's do a deep dive into both these objects.

Snowflake Tables

Snowflake stores data in tables logically structured as relational tables in the form of records (rows) and fields (columns). There are four types of tables in Snowflake as covered in the following sections.

Permanent Tables

These are regular database tables. Consider these as the default table type in Snowflake that is created when users execute the `CREATE TABLE` command (they do not need any additional parameters during table creation). Data stored in permanent tables consumes space and contributes to the storage charges. It also comes with standard features such as Time Travel and Fail-safe (7 days), which helps with data availability and recovery but also results in higher storage costs. Snowflake Time Travel allows users to access historical data at any point in time.

© Ruchi Soni 2023
R. Soni, *Snowflake SnowPro™ Advanced Architect Certification Companion*, Certification Study Companion Series, https://doi.org/10.1007/978-1-4842-9262-4_6

The Fail-safe period starts immediately after the Time Travel retention period ends and provides a 7-day period during which historical data may be recoverable by Snowflake. Figure 6-1 shows the Snowflake continuous data protection (CDP) life cycle. All the features are discussed in detail in Chapter 11 of this book.

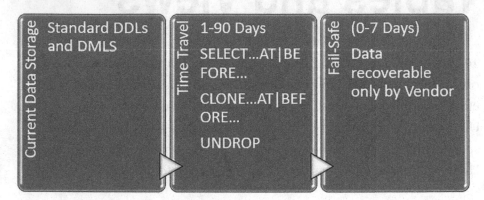

Figure 6-1. CDP in Snowflake

Permanent tables are just like standard tables in any database; they are visible to users with the required privileges until they are dropped by the user. They should ideally be used for use cases with strict data protection requirements.

To create a permanent table, users should simply execute the CREATE TABLE command. Figure 6-2 explains how we can create permanent tables in Snowflake.

```
CREATE OR REPLACE DATABASE SAMPLE_DB;
CREATE OR REPLACE SCHEMA SAMPLE_DB.SAMPLE_TABLES;
CREATE OR REPLACE TABLE EMP_PERMANENT
(FIRST_NAME STRING,
LAST_NAME STRING,
CITY STRING);
```

Figure 6-2. *Creating a permanent table*

Note   The Time Travel period for permanent tables on Snowflake standard edition is 0 or 1 day, and enterprise and higher editions can be from 0 to 90 days. The DATA_RETENTION_TIME_IN_DAYS parameter can be used to explicitly set the retention period (days to retain historical data) for Time Travel. If the value is set to 0, then once the table is dropped, it immediately enters the Fail-safe period.

Transient Tables

Transient tables in Snowflake are permanent tables without a Fail-safe period and a limited Time Travel period (data_retention_time_in_days value as 0 or 1). Here the default retention time is 1 day.

Like a permanent table, transient tables are visible to users with the right privileges until they are dropped. However, transient tables can lose data in the event of a system failure.

To create a transient table, use the TRANSIENT keyword while creating a table. Figure 6-3 explains how we can create a transient table.

```
CREATE OR REPLACE DATABASE SAMPLE_DB;
CREATE OR REPLACE SCHEMA SAMPLE_DB.SAMPLE_TABLES;
CREATE OR REPLACE TRANSIENT TABLE EMP_TRANSIENT
(FIRST_NAME STRING,
LAST_NAME STRING,
CITY STRING);
```

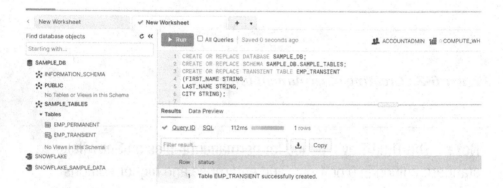

Figure 6-3. *Creating a transient table*

Note   Transient tables cannot be converted to any other type once created.

These tables can be used in cases where they have a lot of data but data protection is not a concern. These tables are also best suited for scenarios where the data in your table is not critical and can be recovered from external means if required. These tables can be used to store data that should be maintained between sessions but can be recovered in the case of any system failure.

Note   Just like permanent tables, transient tables add to the overall storage cost, but since there are no Fail-safe charges, the cost is lower than on permanent tables.

Temporary Tables

Temporary tables are created and maintained by users only for the duration of the session and are not visible in other sessions. Once the session completes, these tables are purged, and data is nonrecoverable. Hence, these tables are used for storing nonpermanent data. These tables are useful for cases where users want to run the same query multiple times, e.g., create different reports on a common dataset. In this case, using a temporary table to store the dataset would be cheaper and faster.

Note   Since a temporary table is purged once a session completes, its Time Travel retention period can be 1 day or the remainder of the session (the smaller value of the two).

To create a temporary table, use the TEMPORARY keyword while creating the table. Figure 6-4 explains how we can create a temporary table.

```
CREATE OR REPLACE DATABASE SAMPLE_DB;
CREATE OR REPLACE SCHEMA SAMPLE_DB.SAMPLE_TABLES;
CREATE OR REPLACE TEMPORARY TABLE EMP_TEMPORARY
(FIRST_NAME STRING,
LAST_NAME STRING,
CITY STRING);
```

Figure 6-4. *Creating a temporary table*

Once created, the temporary table type cannot be changed. Also, since it stores data (for the duration of the session), it contributes to the storage charges while the table is not purged (the session is active).

Note   To avoid storage costs, users should drop temporary tables or exit the session when temporary tables are no longer needed.

External Tables

Unlike the other types of tables mentioned, data in external tables is maintained on external stage, and these tables store metadata information (the name of each staged file, the path of the file, the row number of each record, etc.), as explained in Figure 6-5.

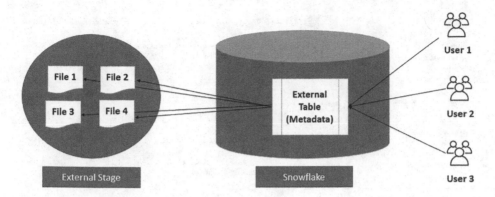

Figure 6-5. *External stage*

Metadata information stored in external tables can be used to access and query data stored in the external stage without moving it into actual Snowflake tables. The data can be used by users to create a data lake by accessing raw data from files, joining it with Snowflake tables, performing join operations, applying complex business logic, and creating views (if required). Since the data is not moved to Snowflake, there is no storage charged within Snowflake. These external tables are useful for cases where organizations want to query data directly from an external stage without the need to duplicate it in Snowflake, which enables users to minimize data silos and maximize the value from their data lakes.

Note   External tables are read-only, so users cannot perform any DML operations.

Figure 6-6 explains how we can create an external table from a file stored in an external stage.

```
create or replace external table employee
with location= @SAMPLE_DB.EXTERNAL_STAGE.SAMPLE_STAGE
file_format= (type = csv field_delimiter=',' skip_header=1)
```

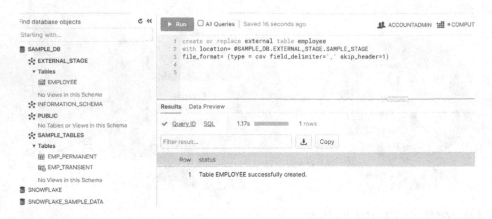

Figure 6-6. *Creating an external table*

Note   Since external tables store data in cloud storage, querying data from external tables is slower than querying database tables directly. However, materialized views (precomputed results) based on external tables can improve query performance. The concept of the materialized view is explained in detail later in this chapter.

External tables include the following columns:

- VALUE: The VARIANT field represents a single record in the file.

- METADATA$FILENAME: This identifies the name and path of each staged file in an external table.

- METADATA$FILE_ROW_NUMBER: This identifies the row number of each staged file record.

Note   Users cannot drop VALUE, METADATA$FILENAME, and METADATA$FILE_ROW_NUMBER columns from an external stage.

Users should use the ALTER EXTERNAL TABLE...REFRESH command to refresh external table metadata. Once executed, this command supports the following activities:

- Adds new files to table metadata

- Updates file changes in table metadata

- Removes files that are no longer available from table metadata

Event notifications for cloud storage can trigger automatic refreshes of the external table metadata. However, the vendor charges 0.06 credits per 1,000 event notifications, so this impacts the overall cost.

Things to Remember

Let's now also understand a few important things that we should be aware of for Snowflake table types.

- These types (permanent/transient/temporary) can also apply to other database objects such as a database or schema. In this case, all the tables created within that type of schema are of the specific type; e.g., if we create a temporary schema, then all the tables within that schema are temporary.

- Snowflake allows users to create the same name for temporary and nontemporary tables in one schema. In this case, the temporary table takes precedence, and all queries performed in the session on the table would only affect the temporary table, which can lead to unexpected behavior.

- The CREATE TABLE privilege on the schema is not required to create a temporary table.

- The temporary tables' retention period completes once it is dropped.

- For data protection, standard tables that are required for the long run should always be defined as permanent to ensure they have an associated Fail-safe period.

- The Fail-safe period is nonconfigurable (7 days).

- One easy way to identify a table as permanent/transient/temporary is by looking at the table icon on the left side of the worksheet.

Figure 6-7 shows the different icons of tables we have created in this chapter.

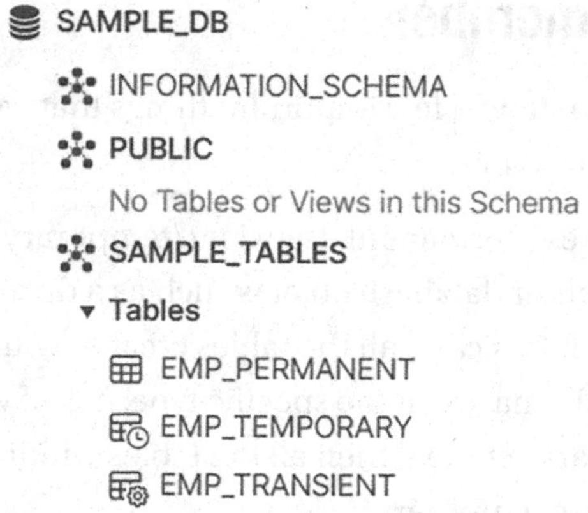

Figure 6-7. Different table icons

CREATE TABLE Command

The CREATE TABLE command is used for table creation. You can create a table from a file, from query results, from a set of staged files, from another table, etc. The following is the simple syntax of this command:

```
CREATE OR REPLACE
<TABLE_TYPE>
 TABLE [ IF NOT EXISTS ] <table_name>
( <col_name> <col_type>,
   <col_name> <col_type>,...)
```

Besides the standard parameters for this command, there are a few optional parameters that users should know, as mentioned here:

- DATA_RETENTION_TIME_IN_DAYS = <n>

  Here <n> is an integer value used to set the retention period (days to retain historical data) for Time Travel for the Time Travel feature. The following are the supported values based on the Snowflake edition and table types:

  - *Standard Edition*: 0 or 1

  - *Enterprise Edition*: Permanent tables (0 to 90)

  - *Enterprise Edition*: Temporary or transient tables (0 or 1)

- MAX_DATA_EXTENSION_TIME_IN_DAYS = <n>

  Here <n> is an integer value that specifies the maximum number of days for which Snowflake can retain data longer than the retention period. The value can be 0 to 90 days, with 0 meaning no extension. The default value is 14 days.

- SKIP_HEADER = <n>

  Here <n> is an integer value that gives the number of lines to skip from the start of the file.

- FIELD_OPTIONALLY_ENCLOSED_BY =
  'character' | NONE

  This option can be used to enclose strings in the
  <character> (single quotes or double quotes) or NONE.

- ERROR_ON_COLUMN_COUNT_MISMATCH = TRUE | FALSE

  If set to TRUE and if the number of delimited columns in
  an input file does not match the number of columns in
  the Snowflake table, it generates a parsing error.

- NULL_IF = ('S1' ,'S2'...)

  This option can be used for string conversion from SQL
  NULL to the first value in the list.

- EMPTY_FIELD_AS_NULL = TRUE | FALSE

  This option specifies whether to insert SQL NULL for
  empty fields.

- STRIP_OUTER_ARRAY = TRUE | FALSE

  For VARIANT data types, if data exceeds 16 MB, then this
  Boolean value allows the JSON parser to remove the
  outer array and load records in separate rows.

- STRIP_NULL_VALUES = TRUE | FALSE

  It instructs the JSON parser to remove (strip) null values
  from the data.

The following are a few points worth mentioning:

- Reserved keywords such as CURRENT_DATE, CURRENT_
  ROLE, CURRENT_TIME, etc., should not be used as column
  identifiers.

- To disable Time Travel for a table, you should set DATA_
  RETENTION_TIME_IN_DAYS as 0.

SQL Syntax

Snowflake has a long list of commands, functions, and parameters available for various operations and supports querying using standard SELECT statements. We'll mention a few that are important from an exam perspective.

Top<n>

Here n refers to the number of rows to return. This command returns the maximum rows specified by the <n> argument.

Please remember to use an ORDER BY clause to control the results returned.

Figure 6-8 shows the result of running Top<n> on the sample CUSTOMER table that is part of the TPCH_SF1 schema. There are a few sample schemas and tables that are available when users create a free Snowflake trial account, and we will try to explain our Snowflake commands using them.

```
Select top 5 c_name from tpch_sf1.customer order by c_acctbal
```

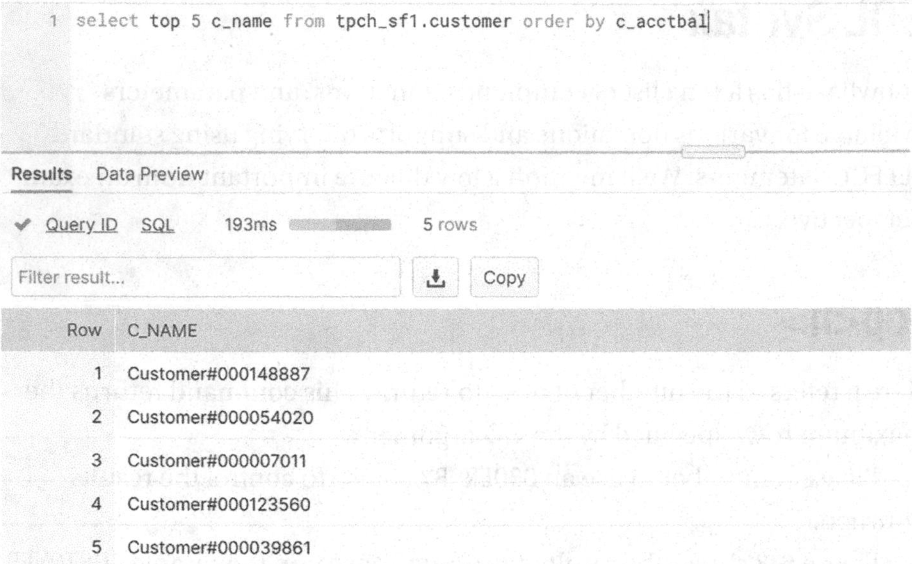

```
1  select top 5 c_name from tpch_sf1.customer order by c_acctbal
```

Results   Data Preview

✔ Query ID   SQL   193ms ▭▭▭▭▭   5 rows

| Filter result... | | 📥 | Copy |

Row	C_NAME
1	Customer#000148887
2	Customer#000054020
3	Customer#000007011
4	Customer#000123560
5	Customer#000039861

Figure 6-8. *Results of top on query*

The previous query gives the top five customers by account balance.

STRIP_NULL_VALUE

This function converts JSON NULL to a SQL NULL value.

Here is the syntax:

STRIP_NULL_VALUE(<col>)

Here, <col> refers to the column name.

Let's run this function on a sample test table that has only one column of the variant datatype. I have captured the result in the column before and after running the function.

Figure 6-9 shows a query explaining the application of the STRIP_ NULL_VALUE function.

Select val:c,strip_null_value(val:c) from myschema.test

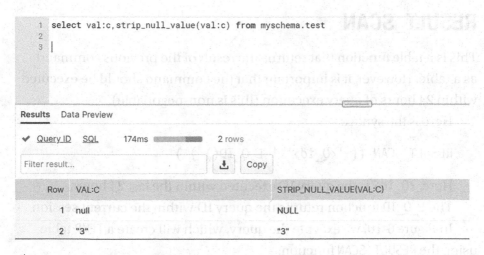

Figure 6-9. *Results of STRIP_NULL_VALUE*

STATEMENT_TIMEOUT_IN_SECONDS

This parameter can be set within the session or within an individual warehouse to control the runtime for SQL statements. As the name suggests, it is used to specify the time (in seconds) after which a SQL statement is canceled by Snowflake. This time is provided in seconds. The default value of this parameter is 48 hours. This parameter can be specified at both the session (account, user, session) and individual object (warehouse) levels (the lowest of the two would take precedence).

Consider an example where users are executing a long-running query on an XL warehouse. Now, as a Snowflake admin, if you want to ensure that a warehouse should not execute any query for more than "x" hours, then you can use this parameter. This will help you save costs and use the proper controls.

RESULT_SCAN

This is a table function that returns the result of the previous command
as a table. However, it is important that the command should be executed
within 24 hours of query execution (this is non-negotiable).

Here is the syntax:

```
RESULT_SCAN ({ '<Q_id>' | L_Q_ID() } )
```

Here, <Q_id> is the query ID executed within the last 24 hours.

The L_Q_ID function returns the query ID within the current session.

In Figure 6-10, we execute the query, which will create a TEMP table
using the RESULT_SCAN function.

```
Create or replace table myschema.temp as
select * from table(result_scan(last_query_id()));
```

```
1  create or replace table myschema.temp as
2  select * from table(result_scan(last_query_id()));
3
4
```

Results   Data Preview

✔ Query ID   SQL   940ms ▓▓▓▓▓▓   1 rows

Filter result...   ⬇ Copy

Row	status
1	Table TEMP successfully created.

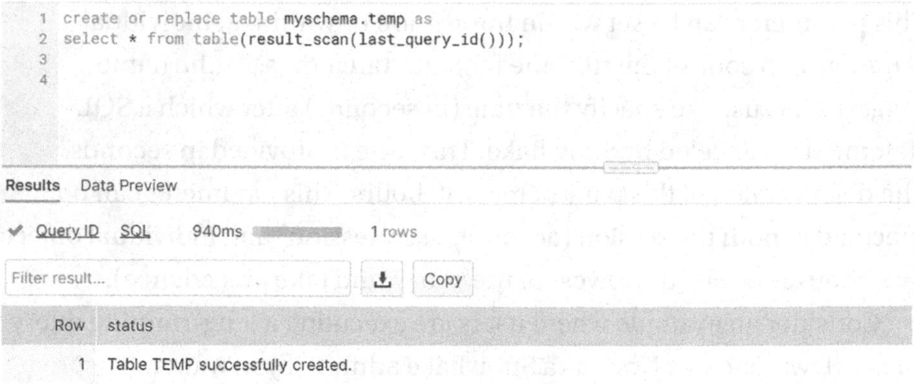

Figure 6-10. *Understanding RESULT_SCAN*

Using Dates and Timestamps

This section describes a few important datatypes supported in Snowflake
to manage the date and time. To store date and time information,

Snowflake provides different datatypes that include DATE (stores dates without timestamp), DATETIME (date and time), TIME, TIMESTAMP, and three variations of TIMESTAMP.

- **TIMESTAMP_LTZ:** This stores UTC time with a defined precision. All operations are performed in the current session's time zone.

- **TIMESTAMP_NTZ:** This stores wall clock time with specified precision. All operations are performed without taking any time zone.

- **TIMESTAMP_TZ:** This stores UTC time with a time zone offset. All operations are performed with a record-specific time zone offset.

Figure 6-11 illustrates the different results for the three variations applied on the same date/time.

```
select '2022-01-02 16:00:00 +00:00' ::timestamp_ltz,
'2022-01-02 16:00:00 +00:00' ::timestamp_ntz,
'2022-01-02 16:00:00 +00:00' ::timestamp_tz
```

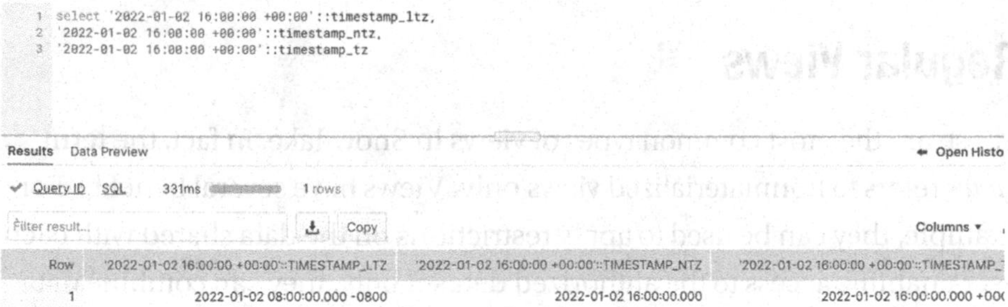

Figure 6-11. *Understanding DATE and TIMESTAMP*

Snowflake Views

Now let's work on the second important topic of this chapter; we will discuss different types of views and related characteristics. People well versed in database terminologies understand that by using views users can access the result of a query like a table. You can consider the view as a virtual table that can be used almost anywhere that a table can be used (joins, subqueries, etc.). Each view has a query associated with them, and whenever users query a view, this SQL query gets dynamically executed. Users can use a CREATE VIEW command to create a view.

Note   If the schema is not specified, then Snowflake assumes that the table is in the same schema as the view.

Snowflake provides three types of views.

- Regular (nonmaterialized) views

- Materialized views

- Secure views

Regular Views

These are the most common types of views in Snowflake. In fact, the term *views* refers to nonmaterialized views only. Views have several benefits. For example, they can be used to apply restrictions on the data shared with each user enabling access to the authorized dataset only, they can combine data from multiple tables and present as a single view, etc. A nonmaterialized view's results are created by executing the query at runtime. Figure 6-12 explains how to create a view called V1 from the table Account.

```
create view v1 as
select Account_Name from Account
```

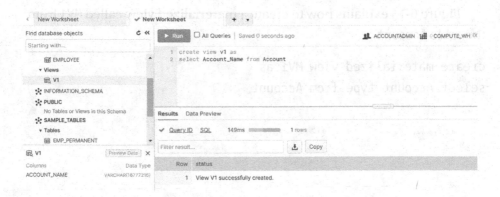

Figure 6-12. *Creating a view*

Since regular view results are not stored for future use (zero data caching), associated SQL queries should be executed again every time a view is accessed. Hence, the performance of regular views is slower than materialized views. However, in some cases, regular views do help Snowflake generate a more efficient query plan.

Materialized Views

A materialized view stores the precomputed results of a view query definition. Their results are stored like a table, which requires storage space and results in storage costs. Materialized views are maintained by Snowflake using compute resources that also result in credit usage.

Since the results of the materialized view are precomputed, querying a materialized view is faster than executing a query against the base table of the view. Repeated complex aggregation and selection operations on large data sets are good use cases for creating materialized views.

Note   Materialized views do not support joins.

Figure 6-13 explains how to create a materialized view called MV1 from the table Account.

```
create materialized view MV1 as
select Account_type from Account
```

Figure 6-13. Creating a materialized view

Materialized views provide the following benefits:

- Better performance of queries that use the same subquery (you can create materialized views on a subquery).

- The maintenance of materialized view is managed automatically by Snowflake, which is faster and easier than manual maintenance. Also, because of this automatic maintenance, materialized views always have access to the latest table data.

- Use the REFRESHED_ON and BEHIND_BY columns in the result of the SHOW MATERIALIZED VIEWS command to know when a materialized view is refreshed.

- Materialized views are shown by INFORMATION_SCHEMA. TABLES and not by INFORMATION_SCHEMA.VIEWS.

- They are more flexible than cached results and faster than tables because they use the "cache."

- Like regular views, materialized views also support data to be hidden at the row and column levels (masking).

Note   If the base table data changes and the query is run before the materialized view is updated, Snowflake either first updates the materialized view or uses updated portions of materialized view and retrieves any new data from the base table. The user is not aware of this because it is managed intelligently by Snowflake.

The following are some of the valid use cases of materialized views:

- A query refers to a small number of rows/columns compared to the base table.

- The results of SQL require significant processing.

- The SQL query is on an external table.

- The data of the base table does not change frequently.

Note   Users should grant privileges to materialized views as they do not inherit the privileges of the base table. If the user wants to access a materialized view, then they need privileges only on the view and not on the underlying objects used in the view SQL.

The following are the limitations of a materialized view:

- They cannot query any other view/user-defined table functions.

- They cannot include UDFs, HAVING, JOIN, ORDER BY, LIMIT, or GROUP BY (keys, grouping sets, rollup, cube).

- Many aggregate functions or DML operations are not allowed in a materialized view.

- Users cannot truncate materialized views.

- If users clone a schema or a database containing a materialized view, the materialized view will be cloned and included in the new schema or database.

- In the case of cloning a schema with a materialized view, the view is cloned in the new schema.

- Time Travel on materialized views is not allowed.

Note   In the case of any addition of columns to the base table, these new columns are not automatically propagated to the materialized views. Similarly, if base columns are dropped, then all materialized views on that base table are suspended, and users must re-create the view again.

Regular vs. Materialized View

From an exam perspective, it is important to understand when we should use regular views versus materialized views. Users should use the following evaluation criteria:

- The following are the use cases for creating materialized views (*all* the following options are true):

  - Results of the view change often.

  - The view is used often.

  - The query is resource intensive.

- The following are use cases for regular views (*any* of the following are true):

  - The results of the view change often.

  - The view is not used often.

  - the query is not resource intensive.

Secure Views

Secure views, as the name suggests, are exposed only to authorized users. They are specifically designed for data privacy and limit access to the data definition of the view so that sensitive data is not exposed to all the users. There can be an impact on query performance for secure views because while evaluating them, the query optimizer bypasses a few optimizations. Figure 6-14 explains how we can create a secure view called SV1 from the Account table.

```
create or replace secure view SV1 as
select * from Account
```

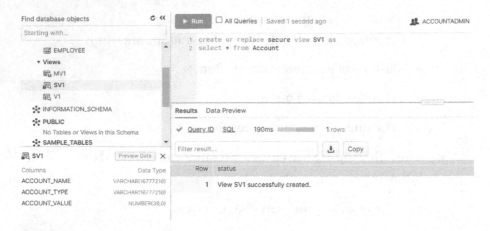

Figure 6-14. *Creating a secure view*

Note   For non-materialized views, the SHOW VIEWS command and for materialized views the SHOW MATERIALIZED VIEWS command are used to identify secure views.

Figure 6-15 explains how we can determine if a materialized view is secure.

```
show materialized views
```

Figure 6-15. *The IS_SECURE option for a materialized view*

Summary

Snowflake provides different options for tables and views. Like any relational data store, data in Snowflake is stored in tables with views, allowing users to grant access to a portion of data in the table. In this chapter, we discussed both tables and views in detail, which includes their advantages/disadvantages and when we should use which object.

CHAPTER 7

Clustering and Micro-Partitions

Traditional databases use static partitioning where users partition tables based on an attribute. In this case, the number of partitions is fixed, but the size may not be the same. This requires users to include a partition clause in a `CREATE TABLE` statement to create a partitioned table. This form of partition has limitations such as maintenance overhead, internal fragmentation, etc.

Snowflake, on the other hand, has a unique way of partitioning called *micro-partitions*, which combine the benefits of static partitioning with additional qualities. In this chapter, we will discuss Snowflake micro-partitions and clustering in detail.

Micro-Partitions

As mentioned in the previous chapter, Snowflake stores data in tables logically structured as records (rows) and fields (columns). Internally, it stores data structured in a columnar fashion as encrypted, compressed files called *micro-partitions*. When users load data, Snowflake automatically compresses and encrypts data and divides it across different micro-partitions. It also maintains metadata of records stored in micro-partitions, which includes information such as the number of distinct values, the range of values for every field, and a few additional details used for optimizing query performance.

R. Soni, *Snowflake SnowPro™ Advanced Architect Certification Companion*, Certification Study Companion Series, https://doi.org/10.1007/978-1-4842-9262-4_7

Each micro-partition can store data between 50 MB to 500 MB (uncompressed), which allows for better pruning and results in better query performance. Within micro-partitions, columns are stored independently (columnar storage), which enables effective scanning and faster query performance.

Micro-partitions are immutable files, which means they cannot be modified once created. These immutable files are stored on the underlying cloud storage. All DML operations take advantage of the underlying micro-partition metadata for table maintenance.

Snowflake stores metadata about these micro-partitions that includes the following:

- Range of values

- Count of distinct values

- Properties for query optimization (including clustering information explained in the next section)

Using an example, let's understand how data is internally stored in Snowflake. Here, *original table* refers to the logical structure of the Snowflake table. Physically this information is stored in the form of micro-partitions. Here we have shown two micro-partitions: Micro-Partition 1 containing data from the first three rows and Micro-Partition 2 with data from the next three rows. See Figure 7-1.

Employee_ID	Employee_Name	City	Salary
11	James	New Orleans	5000
22	Josephine	Brighton	3000
33	Lenna	Anchorage	6000
44	Simona	Ashland	7000
55	Minna	Kulpsville	2000
66	Kiley	Los Angeles	11000

Original Table

11	22	33	44	55	66
James	Josephine	Lenna	Simona	Minna	Kiley
New Orleans	Brighton	Anchorage	Ashland	Kulpsville	Los Angeles
5000	3000	6000	7000	2000	11000

Micro-Partition 1 — Micro-Partition 2

Figure 7-1. *Understanding micro-partitions*

As shown, data is stored internally in columns instead of rows, which enables query pruning. Now let's understand query pruning in detail in the next section.

Query Pruning

Snowflake stores metadata about micro-partitions, which includes a range of values, distinct values, and information for optimization and effective query processing. This metadata information is used to directly identify the micro-partition that contains data corresponding to a user query instead of scanning the entire dataset. This process is known as *query pruning*. Pruning works with semistructured columns in the same way as structured data. For this, the VARIANT datatype is used, which stores semistructured data in its native form.

Let's understand this by running a specific query referring to the same dataset provided earlier. Assume we want to get the names of employees who reside in Ashland. The query to be executed is as follows:

```
Select employee_name
From employee
Where city='Ashland';
```

When this query executes, Snowflake quickly scans the available micro-partitions (based on metadata information) to determine which contains `city = 'Ashland'`. In our example, since there are only two micro-partitions and one of them contains this data, query pruning has reduced data to one micro-partition only. However, a typical Snowflake table may contain millions of micro-partitions (based on the amount of data stored). Additionally, since `employee_name` is required as output, the query will "prune" the micro-partitions that do not contain data for `employee_name`.

Since within micro-partitions data is stored in a columnar format, data scanning for the user query is done in a columnar way, which means if a query filters by only one field, then there is no need to scan an entire partition. Once micro-partitions referred by a query are identified, then only the required columns in the micro-partition are queried, and Snowflake also identifies matching records for the query using this metadata information. This results in faster query execution and optimized performance.

In general, Snowflake follows these guidelines for query pruning:

- Prune micro-partitions not needed by the query.

- Prune columns not needed by the query (within remaining micro-partitions).

In principle, query pruning is effective if the ratio of scanned micro-partitions and data to the ratio of actual data selected is close. The effectiveness of this query pruning can be understood well using the Query Profiler, which is a useful tool provided by Snowflake to analyze the queries. We discussed this Query Profiler in detail in Chapter 15.

> **Note**   As a rule, Snowflake does not prune queries based on a predicate with a subquery.

Clustering

In simple terms, *clustering* means dividing datasets into small groups (clusters) based on data similarity. Data is partitioned to give performance benefits (real-time) enabling queries to focus on only the datasets that are required in the query. Clustering is important for better query performance, particularly with large tables.

Why Clustering?

Data clustering is used by Snowflake for efficient data pruning resulting in optimized query performance. It involves organizing data based on the contents of one or more columns (called *clustering keys*) in the table. During data loading into a table with defined clustering keys, Snowflake automatically sorts data based on these keys.

The whole idea behind data clustering is to improve performance and reduce the cost of query execution. In general, large-size tables are more beneficial for data clustering, and as a guide, if the table size is more than 1 TB, then a user-defined clustering key is recommended. For smaller tables, based on query performance and data size, users can decide whether to go for data clustering, but in this case, there is a high chance that clustering cost might surpass its benefits. As a practice, users can clone the table and apply the clustering approach on the cloned table and check if the query performance improves. If it does, then users can drop the cloned table and make changes to the original table. It is recommended that users should test a few representative queries on the table to set performance baselines before deciding on clustering.

Again, tables that do not change frequently and are queried regularly are good candidates for clustering. In fact, the more frequently a table is queried, the better benefit provided by clustering. Queries benefit from clustering when they filter or sort on the clustering key of the table. However, clustering would be expensive if the table changed very quickly. As a rule of thumb, clustering is best if a table meets *all* the following criteria:

- Large table with multiple terabytes (TB) of data
- Queries filter or sort by clustering keys
- Most of the queries on the table use/benefit from clustering key(s)

Note   Clustering is not beneficial for all Snowflake tables since there are initial costs of clustering data and maintaining table clusters (requires computing resources that consume credits). In general, for small tables less than 1 TB, there is no need to define a clustering key since on such a small table, natural clustering works great.

Reclustering

As the user executes operations on tables (particularly DML on large tables), it may impact existing table clusters. As more and more datasets are added to the table, it can impact clustering since this new data can be written to blocks with key ranges overlapping previously written blocks. Now one option here is for users to perform the manual sorting of rows on key table columns and then insert them again into the table, which can be complex, expensive, and error prone.

Snowflake automates these tasks by defining one or more table columns as a clustering key and managing the key completely in the background without any manual intervention. Snowflake uses these clustering keys to reorganize data to ensure that similar records are moved to the same micro-partition. This DML operation deletes the affected records and re-inserts them, grouped according to the clustering key.

Like clustering, reclustering a table also consumes credits and has associated storage costs. This credit usage is based on the total data size and the size of data to be reclustered. During reclustering, since records of the table are organized based on the defined clustering key, it results in generating new micro-partitions for the table. In fact, adding even a small number of rows to a table can cause the original micro-partition to be marked as deleted and new micro-partitions to be created. These original micro-partitions are purged after Time Travel and Fail-safe as a standard process.

Clustering Keys

As explained earlier, a clustering key is a set of columns in a table used to organize similar data in the same micro-partitions. Users can also provide multiple access paths to data by creating materialized views each with different clustering keys. Based on these keys, Snowflake organizes data during the creation of individual materialized views.

A table with an associated clustering key is called a *clustered table*. These keys can be defined using the CREATE table or the ALTER table. They can also be altered or dropped at any time. There is no default clustering key in Snowflake. If the clustering key is not defined on a table, then clusters are created during data inserts. Both tables and materialized views can be clustered. If users don't define a clustering key on a table, the table will take its data ingestion/loading order as its natural clustering.

There are several benefits of using a clustering key as mentioned here:

- Pruning resulting in fast data scanning

- Better column compression

Note   Once a clustering key is defined on a table, all future maintenance on the rows in the table is performed automatically by Snowflake.

Selecting the right candidates (fields) for a clustering key is very important. To manage costs, generally Snowflake recommends three-fourths of the columns/expressions to be used as clustering keys as the more clustering keys, the bigger the overhead to maintain the clustering order of data. The following is the recommended order for creating cluster keys:

- Columns frequently used in selective filters
- Columns frequently used in joining predicates

Column cardinality (number of distinct values) is another important aspect of deciding the table clustering key. It should have large enough distinct values for effective query pruning and small enough distinct values for co-locating data in the same micro-partitions.

Note   For clustering, using a column with low cardinality might result in minimal pruning, using a column with very high cardinality is more expensive, and using a unique key for clustering might be more costly than beneficial, so users should be careful while deciding on a clustering key (refer to the guidelines provided earlier while choosing clustering keys). In the case of multicolumn clustering, Snowflake recommends ordering columns from lowest to highest cardinality.

SYSTEM$CLUSTERING_DEPTH Function

The SYSTEM$CLUSTERING_DEPTH function is used to calculate the average depth of the table according to the clustering keys specified in the arguments. Its value varies based on the clustering keys.

The CLUSTERING_DEPTH of a table containing data is always greater than or equal to 1. The smaller its value, the better clustered the table for specified columns. Figure 7-2 explains the clustering depth of a table. At the beginning there is overlap between micro-partitions, so the value of clustering depth is more, but as this overlap decreases, the clustering depth value also decreases, and eventually it will become constant.

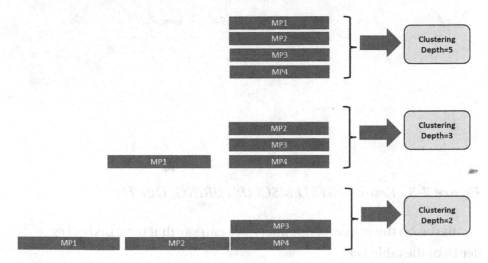

Figure 7-2. *Clustering depth*

The following is the syntax:

```
SYSTEM$CLUSTERING_DEPTH('<t1>' , '( <c1> , <c2> ... )' [ ,
'<p>' ] )
```

The following are the details of the arguments:

```
<t1>:- Table name
(<c1> , <c2> ... ] ) :- Columns list to calculate
clustering depth
<p>: Expression that filters the range of values in the columns
```

Figure 7-3 runs a query on how to calculate SYSTEM$CLUSTERING_DEPTH.

```
SELECT SYSTEM$CLUSTERING_DEPTH('tpcds_sf100tcl.call_
center','cc_call_center_sk')
```

Figure 7-3. *Result of SYSTEM$CLUSTERING_DEPTH*

Based on the results of this query, we can say that the clustering_
depth of the table is 1.

Note   The clustering depth for a table with no micro-partitions is 0.

Please note that its value is not the right measure of the clustering
effectiveness of a table. The best way to know if a table is clustered well
is based on whether queries on a table are performing as required and
whether the performance does not degrade with time.

SYSTEM$CLUSTERING_INFORMATION Function

This function is used to determine how well a table is clustered. It takes in a table name and a list of columns as inputs and outputs clustering information of the table. The following is the function syntax to be used:

```
SYSTEM$CLUSTERING_INFORMATION('<t1>' [ , '( <e1> , <e2>
... )' ] )
```

The following are the details of arguments:

```
<t1>:- Table name
(<e1> , <e2> ... ) :- Column names/expressions to calculate
clustering information
```

It returns a JSON object that provides details of the clustering. This includes the following fields:

- *Cluster_by_keys*: Columns to return clustering information

- *Notes*: Guidance on effectiveness of clustering

- *Total_partition_count*: Total number of micro-partitions created for the table

- *Total_constant_partition_count*: Total number of constant micro-partitions for the table

- *Average_overlaps*: Average number of overlapping micro-partitions

- *Average_depth*: Average overlap depth of each micro-partition in the table

- *Partition_depth_histogram*: Histogram depicting the distribution of overlap depth for each micro-partition in the table

101

Understanding these functions is important from an exam perspective. The high value of `Average_overlaps` or `Average_depth` indicates that the table is not well-clustered.

The query in Figure 7-4 returns the clustering information of the `tpcds_sf100tcl.call_center` table.

```
SELECT SYSTEM$CLUSTERING_INFORMATION('tpcds_sf100tcl.call_
center','cc_call_center_sk')
```

Figure 7-4. *Results of SYSTEM$CLUSTERING_INFORMATION on tpcds_sf100tcl.call_center table*

Based on the results of this query, we can say that this table is clustered by one column (`cc_call_center_sk`). There is only one micro partition (`total_partition_count`), and `average_depth` is 1. Since there is only one micropartition and the clustering depth is equal to the number of partitions in the table, this means a query with a filter on the `cc_call_center_sk` column will always read the entire table, which is not good.

As an example, let's run this function on a Snowflake table called `sample` using three columns.

```
select system$clustering_information('sample', '(c1, c2, c3)');
```

Assume the following is the output:

```
|   "cluster_by_keys" : "(c1, c2,c3)",                              |
|   "total_partition_count" : 200                                  |
|   "total_constant_partition_count" : 2,                          |
|   "average_overlaps" : 102.01,                                   |
|   "average_depth" : 75.02,                                       |
|   "partition_depth_histogram" : {                               |
|     "00000" : 0,                                                 |
|     "00001" : 0,                                                 |
|     "00002" : 2,                                                 |
|     "00003" : 2,                                                 |
|     "00004" : 4,                                                 |
|     "00005" : 5,                                                 |
|     "00006" : 4,                                                 |
|     "00007" : 5,                                                 |
|     "00008" : 7,                                                 |
|     "00009" : 7,                                                 |
|     "00010" : 4,                                                 |
|     "00011" : 60,                                                |
|     "00012" : 100,                                               |
|   }                                                              |
| }
```

Understand from the output that this sample table is not well-clustered because of multiple reasons like the following:

- There are 2 constant partitions out of 200 partitions.

- There is a high value for average_overlaps and average_depth.

- If you see Partition_depth_histogram, then most of the values are grouped in the lower parts.

Summary

In this chapter, we discussed micro-partitions and clustering in Snowflake. Users can improve the query performance by selecting the right clustering keys. Snowflake stores metadata about micro-partitions, which includes a range of values, the number of distinct values, and information about optimization and effective query processing. This metadata information is uscd to directly identify the micro-partition that contains data corresponding to a user query instead of scanning the entire dataset.

CHAPTER 8

Cloning

While working on projects, it might be required to copy data from one environment to another. This is called *database cloning*, which means creating a point-in-time copy of a database. Often the need here is for the developers to test the functionality of code using the current structure and data or for the data scientists to just play around with data. In a traditional database system, it might take days to spin up a copy of production data into another environment and then cost extra for maintaining this data copy. On the other hand, Snowflake provides a convenient way of creating these clones. In this chapter, we will discuss the Snowflake cloning feature in detail.

Zero-Copy Clone

Cloning, often referred to *zero copy cloning*, is a powerful feature in Snowflake that enables users to take a "point-in-time" snapshot of tables, schemas, and databases and generate a reference to an underlining partition that originally shares the underlying storage until users make a change. In other words, it helps to create a copy of data without replicating data, which has several benefits that include no data duplication, less storage cost (a clone of a table refers to the same micro-partition unless some data changes), and quick environment creation.

© Ruchi Soni 2023

R. Soni, *Snowflake SnowPro™ Advanced Architect Certification Companion*,
Certification Study Companion Series, https://doi.org/10.1007/978-1-4842-9262-4_8

Create and Identify a Clone

Snowflake's zero-copy cloning feature ensures data is immediately available for use across users/environments with no additional time/cost. Users can create a clone of a table using the CLONE command. There are multiple real-world use scenarios that benefit from cloning. For example, consider the case when you need to create a copy of one database (1 TB size) for running a few test scenarios. This can be quickly done using the Snowflake zero-copy cloning feature. Since cloned tables share the same underlying storage (unless changed), this can help save money for business. Again, consider a case when the ETL process corrupts data in the underlying table; then the business can use a clone for the table to quickly roll back to the previous state.

Figure 8-1 explains the use of the CLONE command.

CREATE OR REPLACE TABLE EMPLOYEE_V2 CLONE EMPLOYEE

Figure 8-1. *Cloning a table*

Every Snowflake table has an ID used to uniquely identify the table. Similarly, a clone of a table can be identified using the CLONE GROUP ID, which indicates whether the table is cloned. These fields are available in the INFORMATION_SCHEMA.TABLE_STORAGE_METRICS table.

For a table, if the value of ID and CLONE_GROUP_ID is different, the table is cloned; otherwise, it is not. Again, if a table is cloned, then CLONE_GROUP_ID can be used to identify the ID of the table for which it is a clone.

Figure 8-2 illustrates this with an example.

```
SELECT TABLE_NAME, ID, CLONE_GROUP_ID
FROM INFORMATION_SCHEMA.TABLE_STORAGE_METRICS
WHERE TABLE_NAME LIKE 'EMPLOYEE%'
```

Figure 8-2. *How to identify a clone*

How a Clone Is Stored Internally

When you execute the CLONE command, then a snapshot of the table (at the time when the command was executed) is considered. Both the clone object and the original object refer to the same micro-partitions. Once a clone is created, then any change to the original object is not propagated to the clone object, and vice versa. This means that changes in both objects are mutually exclusive.

Let's understand this using an example. In an earlier example, let's assume the EMPLOYEE table is stored in four micro-partitions. Now when users create a clone EMPLOYEE_V2, Snowflake creates a new set of metadata pointing to the identical micro-partitions that store EMPLOYEE data, as shown in Figure 8-3.

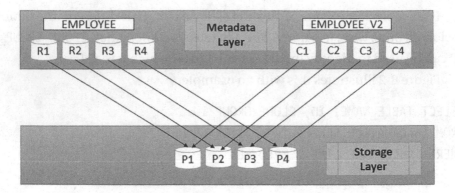

Figure 8-3. *Storage when the clone is created*

Now let's assume users change the data of the clone table EMPLOYEE_
V2, which is part of micro-partition P3. Here Snowflake duplicates a
modified micro-partition P3 and generates a new micro-partition P5 and
assigns it to the stage environment, as depicted in Figure 8-4.

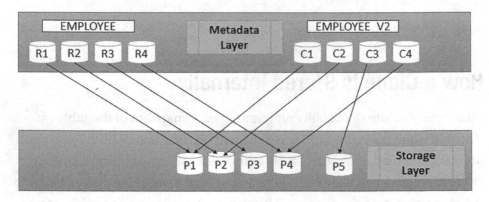

Figure 8-4. *Storage post update to clone*

When created, the clone object inherits the structure, data, and a few
properties from the original object. However, it does not include the source
table load history, which means users can reload the same files in the
clone object as were loaded in the source table.

A clone can be replicated any number of times in Snowflake with each clone object containing shared storage for data that has not changed and independent storage to take care of new data (inserts/updates). For large objects, cloning is not instantaneous and does not lock the object being cloned.

Note   The replication of a clone replicates data too, which increases user data storage cost.

Privileges

The following is the minimum set of privileges required to create a clone:

- SELECT privilege for the source table

- OWNERSHIP privilege for the source objects like pipes, streams, and tasks

- USAGE privilege on other objects

The following are the additional rules that apply:

- When created, the clone inherits the structure, data, and a few properties from the source object.

- If you create a clone of a database or schema, the clone includes all child objects active at the time of creation.

- During cloning, the owner of the clone should explicitly copy privileges from the source object (no automatic propagation). However, if the clone is created for a database/schema, it inherits all the granted privileges on the clones of child objects.

- To copy grants to the clones, users should use the CREATE <object> option since the CREATE CLONE command does not copy grants on the source object to clone.

Other Considerations

There are additional considerations for cloning. Some of them are mentioned in detail here:

- Database, schema, tables, streams, stage, file format, sequence, and tasks can be cloned in Snowflake.

- Snowflake supports cloning permanent/transient/temporary tables. However, temporary tables can be cloned only to a temporary table or a transient table.

- CLONE supports the Time Travel option for database, schema, and tables (except temporary tables). This can be done using the AT | BEFORE clause in the CLONE command. However, objects should exist at the specified time; otherwise, Snowflake throws an error.

- A cloned table references a cloned sequence if the database containing both the table and the sequence is cloned.

- In the case of PK and FK constraints, if both tables are in the same database/schema and the database/schema is cloned, the cloned table with the FK refers to the PK in the cloned table. However, in the case of a separate database/schema, the cloned table with an FK refers to the PK of the source table.

- If the object cloned has a clustering key defined, then the clone also has the same clustering key. However, auto-clustering for a clone is suspended by default.

Note   When cloning a database, external named stages in the source are cloned. However, internal named stages/pipes and external tables are *not* cloned.

- When a database/schema is cloned, the following happens:

  - Pipes that reference an internal stage are not cloned and that reference an external stage are cloned.

  - Records in the stream that are not consumed (in the clone) are no longer inaccessible.

  - Tasks are suspended by default and should be resumed manually in the clone.

  - Any child object that did not exist at the specified point (in the case of Time Travel is not cloned).

- The state (default) of a cloned pipe is as follows:

  - PAUSED when AUTO_INGEST = FALSE

  - STOPPED_CLONED when AUTO_INGEST = TRUE

- As a best practice, users should not execute any DML commands on the source object while the cloning operation is running, since cloning is not instantaneous and does not lock the source object while cloning. Another option is to set DATA_RETENTION_TIME_IN_ DAYS=1 for the tables considered for cloning and reset it to 0 once cloning completes.

Summary

There are multiple reasons for cloning data stores. Snowflake offers zero copy cloning as a very powerful feature that provides a quick and easy way to copy an object (table/schema/database) without incurring any additional costs.

In this chapter, we discussed different aspects of the Snowflake feature called zero copy cloning. It duplicates the structure, data, and a few other parameters of the original table. When a table is cloned in Snowflake, both the clone object and the original object refer to the same micro-partitions. Once a clone is created, then any change to the original object (addition/deletion/update) is not propagated to the clone object, and vice versa; both source and clone are independent of each other. Zero-copy clones create a point-in-time copy of a database very quickly, which helps users save money, effort, and time.

CHAPTER 9

Secure Data Sharing

Another powerful and easy-to-use feature provided by Snowflake is called *secure data sharing*. With this feature, Snowflake enables the sharing of read-only database tables, views, and user-defined functions (UDFs) from providers to consumers using a concept called *secure shares*. This feature uses the Snowflake services layer and metadata information to share data without moving it from producer to consumer. Now since there is no actual data movement, there is no storage cost involved to read from the shared database; consumers incur compute costs only for querying the data.

What Is a Share?

Theoretically speaking, shares are named objects in Snowflake. They do not have any data but contain information to enable object sharing between accounts. When a database is created from a share, consumers can access all shared objects. Typically, a share contains the following information:

- Access privileges for objects that are shared
- Consumer details (accounts that consume data from share)

Before proceeding with how secure sharing works, first let's understand the different personas involved in secure data sharing.

R. Soni, *Snowflake SnowPro™ Advanced Architect Certification Companion*, Certification Study Companion Series, https://doi.org/10.1007/978-1-4842-9262-4_9

Data Sharing Personas

Secure data sharing includes two types of accounts: data provider and data consumer. As the name suggests, the data provider creates a data share and shares it with other Snowflake accounts (consumers). It uses grants for granular access control to share objects.

Note   Data providers can create unlimited shares and share them with an unlimited number of consumers and vice versa (consumers can consume multiple shares). However, only one database per share can be created.

A data consumer is an account that creates a database from a share to consume data provided by a data provider. Once the shared database is added, consumers can query shared objects like any other normal database (no additional syntax needed). Snowflake provides two types of consumer accounts.

- *Standard consumer accounts*: These are for existing Snowflake customers. In this case, data can be shared directly to the consumer's existing account, and the consumer pays for all compute resources for querying the shared databases.

- *Reader account*: Reader accounts are used for data providers to share data with users who are not Snowflake customers. There is no licensing agreement or associated setup charge involved. Once a reader account is created, it is managed by the data provider, which includes any associated charges for computing resources for querying the shared databases. These accounts can be used only for data queries (additions and updates not supported).

Figure 9-1 and Figure 9-2 give pictorial representations for these two personas.

Note   Querying data is supported only for users with Reader accounts (no DML operations allowed).

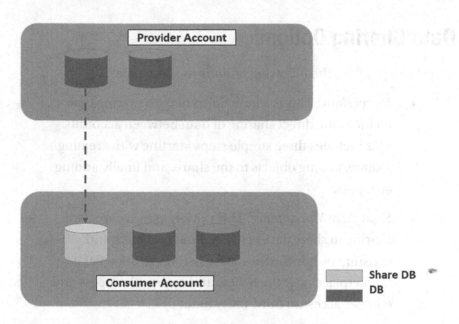

Figure 9-1. *Secure data sharing with Provider and Consumer accounts*

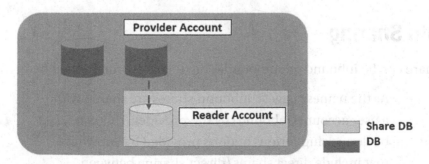

Figure 9-2. *Secure data sharing with Provider and Reader accounts*

115

Note   Once a Reader account is created, it should be added to shares to enable access to objects in the share for the account. Once data sharing is complete, the provider should drop the Reader account to restrict access.

Data Sharing Options

Snowflake provides the following options to share data:

- *Direct share*: This is a basic form of data sharing that includes the direct sharing of data between accounts. This includes three simple steps starting with creating a share, adding objects to the share, and finally adding accounts.

- *Snowflake Marketplace*: This option uses secure data sharing to share data between data providers and consumers. It includes adding listings to Snowflake Marketplace after the review and approval process and enables access to third-party data.

- *Data exchange*: This refers to a user's own data hub for data sharing between a few selected members.

Data Sharing

A share can be inbound or outbound. The details are mentioned here:

- As the names suggest, inbound shares are shared with a user account (inbound) by the provider. This can be done using various options provided by Snowflake that include direct shares (direct sharing between accounts), a data exchange (user data hub for large-scale data sharing), or a data marketplace (Snowflake Marketplace for third-party data access). For an inbound share, users can view all shares from providers and create a database from a share. Inbound requests are requests from data consumers to access your data. Users have the option of approving or denying the request.

- Outbound shares are used to share data with consumers from your account (outbound). Like inbound shares, users can share data via a direct share, a data exchange, or the Snowflake Marketplace. For an outbound share, users can view the shares they have created or have privileges to access/create/edit/revoke access to the share. Outbound requests are requests that users have submitted for data listings from other providers in the data exchange. You can sort the requests by status. If a request has been denied, a comment is provided next to the request.

Data Sharing Steps

The following are the steps required to enable data sharing in Snowflake:

1. Create database roles.

2. Grant privileges on objects to database roles.

3. Create an empty share.

4. Add a database to a share (grant privileges for a database and objects to the share).

5. Add accounts to a share.

117

Managing Shares

Users can manage shares either using the web interface or using
SQL commands. For the exam, let's understand a few important SQL
commands associated with using shares.

- SHOW SHARES: This provides a list of available shares
  including outbound shares (to consumers) created
  in the user account (as a data provider) and inbound
  shares (from providers) available for consumption.

  Figure 9-3 explains all the shares created or available
  for consumption in my sample account.

  SHOW SHARES;

Figure 9-3. *SHOW SHARES*

From the result of this command, the <u>KIND</u> column displays the type
of share (inbound/outbound). In the case of an outbound share, the <u>TO</u>
column displays the accounts added.

Note   The SHOW SHARES command does not require an active
warehouse.

- DESCRIBE SHARE: It describes the data objects that
  can be included in the share. Figure 9-4 explains this
  command by running it on a sample share.

```
DESCRIBE SHARE SNOWFLAKE.ACCOUNT_USAGE
```

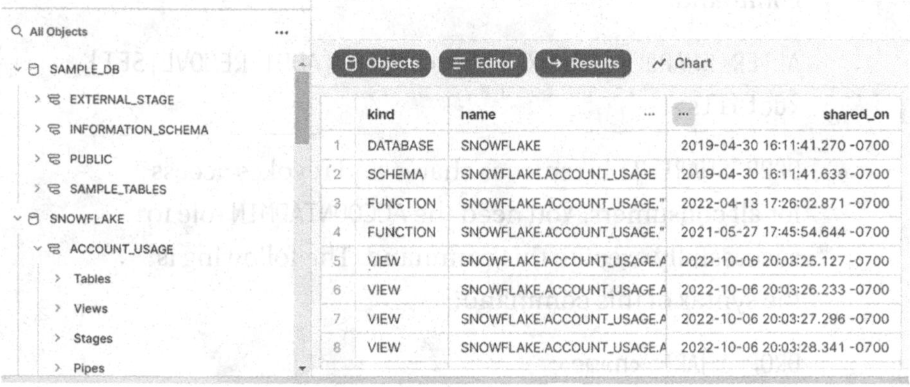

Figure 9-4. *DESCRIBE SHARE*

The KIND column here displays the type of objects in the share.

- CREATE SHARE: This creates a new empty share. Once
  it is created, users can include a database and objects
  in the share using the GRANT <privilege> TO SHARE
  command and can add columns in the share using the
  ALTER SHARE command. The following syntax is used
  for this command:

```
CREATE [ OR REPLACE] SHARE <name>
```

Note   Users can use IMPORTED PRIVILEGES to grant access to other roles and limit access. This is useful in cases where users need to provide read access to only a few objects instead of all the objects in the share.

- ALTER SHARE: This modifies the properties of an
  existing share. This includes adding/removing
  accounts and updating other related properties. You
  need either the OWNERSHIP or CREATE SHARE privilege
  to alter a share. The following is the syntax of this
  command:

  ALTER SHARE [IF EXISTS] <name> {ADD| REMOVE|SET}
  <details>

- DROP SHARE: This removes shares and revokes access
  for all consumers. You need the ACCOUNTADMIN role to
  successfully execute this command. The following is
  the syntax of this command:

  DROP SHARE <name>

Note   Dropped shares cannot be recovered but must be created and
configured again.

Key Considerations for Data Sharing

Shared databases have a few important characteristics and limitations that
are important for exams. They are mentioned here:

- The objects shared between accounts are always read-
  only, which means consumers can view/query data but
  cannot run any DDL/DML.

- Cloning and Time-Travel of shared databases, schemas,
  and tables are *not* supported.

- Modifying the comments for a share is *not* supported.

- Re-sharing with other accounts is *not* supported.

- A data share can be consumed once per account.

- Tables, external tables, secure views, secure materialized views, and secure UDFs can be shared.

- Once a share is created, then the new objects added are immediately available for consumption.

- Users can immediately revoke access to a share.

- Secure views can be used to share data that resides in different databases.

- To ensure strict data control, users can use secure views, secure materialized views, and secure UDFs.

- SHOW SHARES provides a list of available shares for the users.

- SHOW GRANTS OF SHARE lists all the accounts that created a database from the share.

Cross-Region Data Sharing

Data sharing is used to securely share data with consumers in the same cloud/region. Snowflake uses database replication to allow data providers to securely share data with data consumers across regions/clouds. This is supported for all the cloud platforms supported by Snowflake, which includes AWS, GCP, and Azure. It is discussed in detail in subsequent chapters (Chapter 13).

Summary

Secure data sharing is used by Snowflake for data sharing in the same account and region. It provides a network of data providers and Consumer and Reader accounts for data sharing. Database replication is another Snowflake feature for data sharing across regions and clouds. In this chapter, we discussed secure data sharing in detail and briefly touched on cross-region data sharing (data replication). Both are very powerful features provided by Snowflake for users to share data across their business ecosystems.

CHAPTER 10

Semistructured Data

Data is an invaluable asset for any organization that wants to succeed in the modern world. Businesses visualize data as the new oil, but it is valuable only in a usable form. Users want to see its fundamental value and extract useful information from data.

With the advent of big data, four Vs characterize data complexity. This includes volume, velocity, variety, and veracity. Additionally, there are three types of data: structured, semistructured, and unstructured data.

Semistructured data does not contain a defined structure, so analyzing it using traditional methods is difficult due to additional complexity. Also, combining structured with semistructured data is difficult. In this chapter, we will discuss how Snowflake handles semistructured data.

Semistructured Data

Semistructured data does not have a fixed schema, which means that new attributes can be added at any time. The data is organized hierarchically in a nested structure such as in arrays or objects.

Snowflake supports datatypes, which enables businesses to query semistructured data the way we query relational databases, which also includes support for importing/exporting data. Depending upon the structure of the data, the size of the data, and the way that the user chooses to import the data, semistructured data can be stored in a single column or split into multiple columns.

© Ruchi Soni 2023
R. Soni, *Snowflake SnowPro™ Advanced Architect Certification Companion*,
Certification Study Companion Series, https://doi.org/10.1007/978-1-4842-9262-4_10

Snowflake provides three semistructured datatypes that include `VARIANT, OBJECT,` and `ARRAY`.

- A `VARIANT` can hold a value of any other datatype.

- An `ARRAY` or `OBJECT` holds a value of type `VARIANT`.

How Snowflake Handles Semistructured Data

Snowflake provides a popular datatype called `VARIANT` to manage semistructured data. This datatype can store data of any other type, which also includes `OBJECT/ARRAY` with a maximum compressed size of 16 MB. You can consider it as a regular column within a Snowflake relational table.

On the other hand, the `ARRAY` datatype is a list-like indexed datatype that consists of variant values, and the `OBJECT` datatype consists of key-value pairs, where the key is a not-null string, and the value is the `VARIANT` type data. From an exam perspective, it is important to understand how queries are executed on semistructured data.

For better performance, Snowflake stores data in columnar binary format irrespective of the datatype. This is transparent to users, which means users can run queries on Snowflake-like standard relational tables.

VARIANT Data Type

A `VARIANT` datatype can store different types of data, which means a value of any datatype can be implicitly cast to a `VARIANT` value.

Additionally, date and timestamps are stored as strings in a `VARIANT` column.

Figure 10-1 creates a table named `WEATHER_DATA` with only one column with the `VARIANT` datatype.

```
Create table weather_data (val variant);
```

Figure 10-1. *Creating a table for variant data*

Snowflake provides functions for type casting, handling NULLs, etc. A function worth mentioning here is called FLATTEN, which flattens nested values into separate columns that can be used to filter query results in a WHERE clause.

To load the table created, we will create a stage from an AWS S3 bucket, as explained in Figure 10-2.

```
create stage weather_nyc url = 's3://snowflake-workshop-lab/
weather-nyc';
```

Figure 10-2. *Creating a stage*

Now let's list the contents of this stage. This can be done using the `list` command, as shown in Figure 10-3.

```
list @weather_nyc;
```

Figure 10-3. *List contents of a stage*

Now we will load data from the AWS S3 bucket into the `WEATHER_DATA` table by specifying a `file_format` as JSON, as explained in Figure 10-4.

```
copy into weather_data
from @weather_nyc
file_format = (type=json);
```

Figure 10-4. *Loading data into a table*

Let's execute SELECT on the WEATHER_DATA table to check its contents, as shown in Figure 10-5.

```
select * from weather_data limit 10;
```

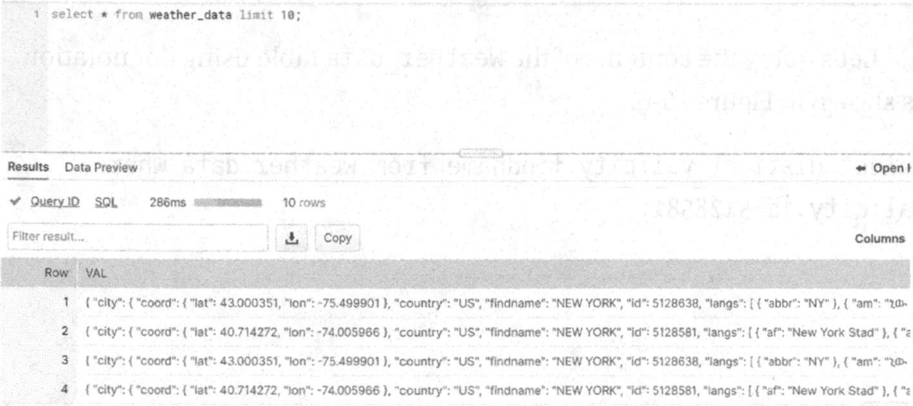

Figure 10-5. *Selecting data from a table*

Query Semistructured Data

Users can query JSON objects using dot notation and bracket notation. The following is the syntax used:

This is the dot notation syntax object: <C1>:<L11_element>.<L2_element>.<L3_element>

This is the bracket notation syntax:

```
<C1>['<L1_element>']['<L2_element>'] ['<L3_element>']
```

Note   It is important to remember that in both the notations, the column name is case-insensitive, and the element names are case-sensitive. For example, in the following list, the first two paths are equivalent, but the third is not:

---

src:employee.name

SRC:employee.name

SRC:Employee.Name

Let's query the contents of the `Weather_data` table using dot notation, as shown in Figure 10-6.

```
select distinct val:city.findname from weather_data where
val:city.id=5128581;
```

Figure 10-6. *Querying semistructured data*

Note   The data size for loading into a VARIANT column is 16 MB (compressed).

Create a View

Users can query the semistructured data and then create views on top of it for data access in a visible and structured manner. Figure 10-7 explains how we can create a view on the `WEATHER_DATA` table.

```
create view weather_view as
select
val:time::timestamp as time,
val:city.id::int as city_id,
val:city.name::string as city_name,
val:city.country::string as country,
val:wind.speed as speed
from weather_data where val:city.id=5128581;
```

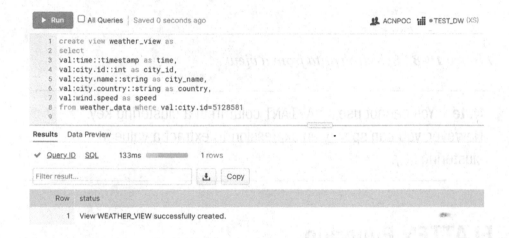

Figure 10-7. *Creating a view*

This view contains five fields and can be queried like a relational view (without any dot or bracket annotation now).

Figure 10-8 shows how you can access data from this view.

```
select * from weather_view;
```

Figure 10-8. *Selecting data from a view*

Note   You cannot use a VARIANT column in a clustering key. However, you can specify an expression to extract a value in a clustering key.

FLATTEN Function

FLATTEN is a Snowflake table function that flattens semistructured data and produces a lateral view of VARIANT/OBJECT/ARRAY fields. It returns a row for each object and can be used to convert semistructured data into a relational format. The following is the syntax used:

```
FLATTEN(INPUT=><expr>[,PATH=><constant_expr>]
[,OUTER=>TRUE|FALSE]
[,RECURSIVE=>TRUE|FALSE]
[,MODE=>'OBJECT'|'ARRAY'|'BOTH'])
```

INPUT is a required parameter, and others are optional.

To understand how the FLATTEN command works, let's run a SELECT on the table WEATHER_DATA, as shown in Figure 10-9.

```
select * from weather_data limit 8;
```

Figure 10-9. *SELECT on WEATHER_DATA*

Now let's explain the FLATTEN function, as demonstrated in Figure 10-10.

```
select val,value
from
   weather_data
  ,lateral flatten(input=>val:city)
 limit 5;
```

Figure 10-10. *FLATTEN function*

In the previous example, executing the FLATTEN command on CITY generates a separate row for every nonarray field in CITY.

Now we will go one level down to see the values of the LANG array using the following command, as shown in Figure 10-11:

```
select val,value
from
   weather_data
   ,lateral flatten(input=>val:city.langs)
limit 5;
```

```
1  select val,value
2  from
3     weather_data
4     ,lateral flatten(input=>val:city.langs)
5  limit 5;
6
```

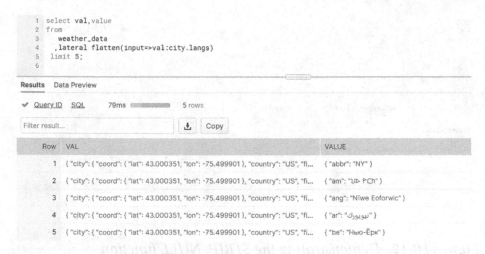

Results   Data Preview

✔ Query ID   SQL   79ms   5 rows

Filter result...		📥 Copy

Row	VAL	VALUE
1	{ "city": { "coord": { "lat": 43.000351, "lon": -75.499901 }, "country": "US", "fi...	{ "abbr": "NY" }
2	{ "city": { "coord": { "lat": 43.000351, "lon": -75.499901 }, "country": "US", "fi...	{ "am": "ኒው ዮርካ" }
3	{ "city": { "coord": { "lat": 43.000351, "lon": -75.499901 }, "country": "US", "fi...	{ "ang": "Nīwe Eoforwīc" }
4	{ "city": { "coord": { "lat": 43.000351, "lon": -75.499901 }, "country": "US", "fi...	{ "ar": "نيويورك" }
5	{ "city": { "coord": { "lat": 43.000351, "lon": -75.499901 }, "country": "US", "fi...	{ "be": "Нью-Ёрк" }

Figure 10-11. Applying FLATTEN one level down

Note   JSON and PARQUET files cannot be loaded into columns in the same table.

STRIP_NULL_VALUE Function

This function internally converts a NULL value from JSON "null" to SQL NULL without any impact on other variant values. We'll demonstrate it using an example, as shown in Figure 10-12.

```
create table test_sample(val variant);
insert into test_Sample
  select parse_json(column1)
  from values
  ('{"var1": "1", "var2": "2", "var3": null}'),
('{"var1": "1","var2": "2","var3": "3"}');
select strip_null_value(val:var3) from test_sample;
```

Figure 10-12. Demonstrating the STRIP_NULL function

Note   Cloud services can help in pruning even if the columns are VARIANT columns.

SYSTEM$EXPLAIN_JSON_TO_TEXT Function

This function returns a VARCHAR containing the EXPLAIN output as text that has been formatted to be relatively easy for humans to read.

The following is the syntax of this function:

```
SYSTEM$EXPLAIN_JSON_TO_TEXT( < format> )
```

Here <format> is an expression that evaluates to a string, containing the EXPLAIN output as a JSON-compatible string.

> **Note**   After a user has stored JSON in a table, you can pass it to the SYSTEM$EXPLAIN_JSON_TO_TEXT function to convert it to a human-readable text format by calling SYSTEM$EXPLAIN_JSON_TO_TEXT.

PARSE_JSON Function

This function parses JSON data, producing a VARIANT value. The following is the syntax for this function:

PARSE_JSON(<expr>)

Here <expr> holds valid JSON information. It returns a value of type VARIANT that contains a JSON document.

> **Note**   The output of this function is NULL if the input is an empty string or a string with only whitespace characters or NULL.

OBJECT_CONTRUCT Function

This function returns an object constructed from the arguments.

OBJECT_CONSTRUCT([<key1>, <value1> [, <keyN>, <valueN> ...]])

> **Note**   If the key or value is NULL, the key-value pair is omitted from the resulting object.

GET Function

This function accepts a VARIANT, OBJECT, or ARRAY value as the first argument and extracts the VARIANT value using the path name as the second argument.

Summary

Snowflake provides support for managing semistructured data by providing native datatypes (ARRAY, OBJECT, and VARIANT) for data storage. It provides the FLATTEN function to convert nested objects into a relational table or store them in native format within the VARIANT datatype. Users can use special operators and functions to query complex hierarchical data stored in a VARIANT column. Regardless of how the hierarchy is constructed, Snowflake converts data to an optimized internal storage format that supports fast and efficient SQL querying.

References

This chapter contains a few references from my other book that covers the basics of semistructured data: *Snowflake Essentials: Getting Started with Big Data in the Cloud* (ISBN 9781484273159).

CHAPTER 11

Time Travel and Fail-Safe

Data is an asset that has a major impact on the long-term success of an enterprise. Access to the right data at the right time is extremely important for businesses to make the right decisions. Two other important features that make Snowflake extremely popular are Time Travel and Fail-safe. Consider an example where data got incorrectly deleted from production systems due to a wrong command/process. In traditional systems, the way to correct this is either to restore from backup or to run the entire data pipeline for a specific table again. However, in Snowflake this can be done in minutes using the Time Travel and Fail-safe options. Snowflake Time Travel enables access to historical data at any point within a defined period.

Time Travel

Snowflake Time Travel allows users to access historical data (that is, data that has been updated or removed) at any point in time. Some of the key benefits include the following:

- Restoring objects that were removed from the database

- Backing up data from critical points

- Analyzing data usage from critical points

© Ruchi Soni 2023
R. Soni, *Snowflake SnowPro™ Advanced Architect Certification Companion*,
Certification Study Companion Series, https://doi.org/10.1007/978-1-4842-9262-4_11

Continuous Data Protection

As the name suggests, *continuous data protection* (CDP) includes a set of features that helps protect data against human errors and software failures. It is designed to provide long-term protection for client data.

Time Travel is a Snowflake feature that helps in the maintenance of historical data. Users can perform the following activities within the specified time using Time Travel:

- Create clones of objects from defined points

- Query data that was accidentally deleted

- Restore objects that were accidently dropped

Figure 11-1 explains the different aspects of continuous data protection that we will discuss in detail in this chapter.

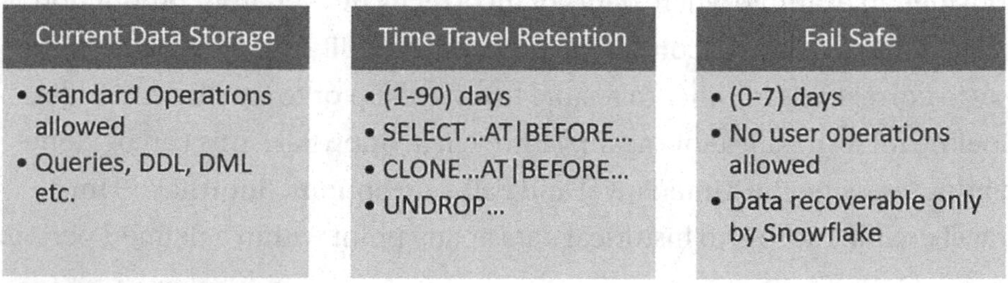

Current Data Storage	Time Travel Retention	Fail Safe
• Standard Operations allowed • Queries, DDL, DML etc.	• (1-90) days • SELECT...AT|BEFORE... • CLONE...AT|BEFORE... • UNDROP...	• (0-7) days • No user operations allowed • Data recoverable only by Snowflake

Figure 11-1. *CDP explained*

Now I'll explain how Time Travel works using an example. Suppose we have an `Employee` table with data shown in Figure 11-2.

```
SELECT * FROM EMPLOYEE;
```

```
1 SELECT * FROM EMPLOYEE;
```

Results   Data Preview

❮ Query ID   SQL   99 rows

[Filter result...]   [↧]   [Copy]

Row	FIRST_NAME	LAST_NAME	CITY	COUNTY	STATE	ZIP
1	James	Butt	New Orleans	Orleans	LA	70116
2	Josephine	Darakjy	Brighton	Livingston	MI	48116
3	Art	Venere	Bridgeport	Gloucester	NJ	8014
4	Lenna	Paprocki	Anchorage	Anchorage	AK	99501
5	Donette	Foller	Hamilton	Butler	OH	45011
6	Simona	Morasca	Ashland	Ashland	OH	44805
7	Mitsue	Toliner	Chicago	Cook	IL	60632

Figure 11-2. *Contents of the Employee table*

Now suppose by mistake you updated the city of the entire table to Chicago. This is shown in Figure 11-3.

```
UPDATE EMPLOYEE SET CITY='Chicago';
SELECT * FROM EMPLOYEE;
```

```
1 UPDATE EMPLOYEE SET CITY='Chicago';
2 SELECT * FROM EMPLOYEE;
```

Results   Data Preview

✓ Query ID   SQL   208ms   ▬▬▬▬   99 rows

[Filter result...]   [↧]   [Copy]

Row	FIRST_NAME	LAST_NAME	CITY	COUNTY	STATE	ZIP
1	James	Butt	Chicago	Orleans	LA	70116
2	Josephine	Darakjy	Chicago	Livingston	MI	48116
3	Art	Venere	Chicago	Gloucester	NJ	8014
4	Lenna	Paprocki	Chicago	Anchorage	AK	99501
5	Donette	Foller	Chicago	Butler	OH	45011
6	Simona	Morasca	Chicago	Ashland	OH	44805
7	Mitsue	Toliner	Chicago	Cook	IL	60632
8	Leota	Dilliard	Chicago	Santa Clara	CA	95111

Figure 11-3. *Updating the city for the entire table*

To fix this, we will use the Time Travel feature provided by Snowflake. Let's use SQL commands to travel 5 minutes back on the table, as shown in Figure 11-4.

```
SELECT * FROM EMPLOYEE AT(OFFSET=> -60*5);
```

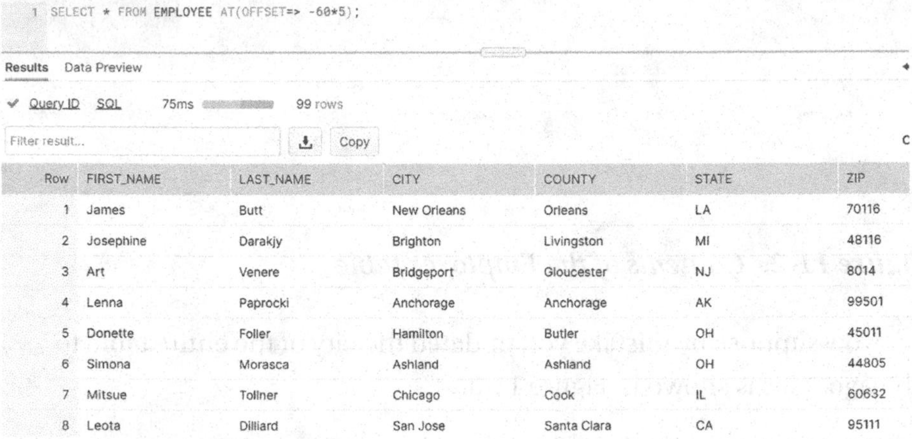

Row	FIRST_NAME	LAST_NAME	CITY	COUNTY	STATE	ZIP
1	James	Butt	New Orleans	Orleans	LA	70116
2	Josephine	Darakjy	Brighton	Livingston	MI	48116
3	Art	Venere	Bridgeport	Gloucester	NJ	8014
4	Lenna	Paprocki	Anchorage	Anchorage	AK	99501
5	Donette	Foller	Hamilton	Butler	OH	45011
6	Simona	Morasca	Ashland	Ashland	OH	44805
7	Mitsue	Tollner	Chicago	Cook	IL	60632
8	Leota	Dilliard	San Jose	Santa Clara	CA	95111

Figure 11-4. *Extracting details from 5 minutes ago*

Now let's use Time Travel to create the original table, as shown in Figure 11-5.

```
CREATE OR REPLACE TABLE EMPLOYEE
AS
SELECT * FROM EMPLOYEE AT(OFFSET=> -60*5);
```

Figure 11-5. *Creating an original table*

Now let's check the contents of the Employee table after running the previous command. This is shown in Figure 11-6.

```
SELECT * FROM EMPLOYEE;
```

Row	FIRST_NAME	LAST_NAME	CITY	COUNTY	STATE	ZIP
1	James	Butt	New Orleans	Orleans	LA	70116
2	Josephine	Darakjy	Brighton	Livingston	MI	48116
3	Art	Venere	Bridgeport	Gloucester	NJ	8014
4	Lenna	Paprocki	Anchorage	Anchorage	AK	99501
5	Donette	Foller	Hamilton	Butler	OH	45011
6	Simona	Morasca	Ashland	Ashland	OH	44805
7	Mitsue	Tollner	Chicago	Cook	IL	60632
8	Leota	Dilliard	San Jose	Santa Clara	CA	95111

Figure 11-6. *Contents of a table*

Hence, the Snowflake Time Travel option enables users to recover data that was accidentally updated, modified, or deleted in a matter of minutes. This is a powerful feature of Snowflake.

Note   When the Time Travel retention period of an object completes, the Fail-safe period starts. While in Fail-safe mode, none of the previous actions can be performed by the user. Instead, the user should raise a ticket with the vendor to restore data.

How to Enable Time-Travel on a Table

Time Tavel can be enabled for a table using the following SQL extensions:

- The AT | BEFORE clause can be included in SELECT (as demonstrated earlier) and CREATE CLONE statements. Additionally, users should use the TIMESTAMP, OFFSET, and STATEMENT options to mention the historical data for Time Travel.

- The UNDROP command can be used for tables, schemas, and databases.

Every query executed is allocated a query ID, which can be used with the Statement option. This query ID can be extracted from the History tab, as shown in Figure 11-7.

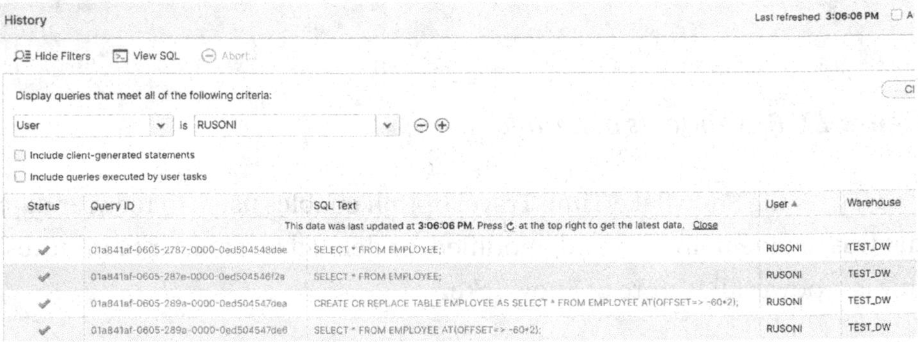

Figure 11-7. *Query ID from the History tab*

We already demonstrated how users can use the Offset parameter for Time Travel. Now Figure 11-8 explains how the Statement parameter can be used.

```
CREATE OR REPLACE TABLE EMPLOYEE
AS
SELECT * FROM EMPLOYEE BEFORE(STATEMENT=>'01a841
af-0605-287e-0000-0ed504546f2a');
```

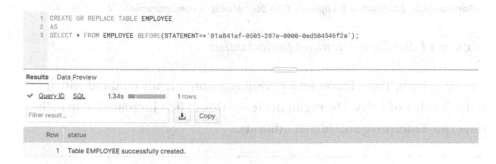

Figure 11-8. *Time-Travel using the Statement parameter*

Note   Users should be careful while executing any Time Travel query. For example, when a Time Travel query is executed on temporary/transient tables, their purging is deferred until the query is running.

Data Retention Period

In simple words, the *data retention* period refers to the time period you want to retain the information. In other words, Snowflake preserves table data for the defined data retention period. Based on the defined data retention period (a number of days), when users run any operation on the table that includes update, delete, or drop, the historical data is retained.

From an exam perspective, it is important to understand the value of the retention period based on the Snowflake edition and table type. This is depicted in Figure 11-9.

Type	Edition	Time Travel Retention Period (Days)	Fail-safe Period (Days)
Temporary	NA	0 or 1(default is 1)	0
Transient	NA	0 or 1(default is 1)	0
Permanent	Standard	0 or 1(default is 1)	7
Permanent	Enterprise & Higher	0 to 90(default is configurable)	7

Figure 11-9. *Data retention period value*

By default, Time Travel for a period is automatically enabled with a default value of 1 day. Users can decrease this to 0 or increase it (up to 90 days) for Enterprise and higher editions.

Note   In a permanent table, if Time Travel is set to 0, it immediately enters Fail-safe mode once the table is dropped.

DATA_RETENTION_TIME_IN_DAYS

As mentioned earlier, this parameter is used to set the retention period (override the default value) for an account to enable Time Travel. This value is defined in days and is used to retain historical data for the Snowflake objects in the database, schema, or individual table. Its value can be as follows:

- 0 or 1 for Standard edition

- 0 to 90 for Enterprise or higher edition

Note   The default value of DATA_RETENTION_TIME_IN_DAYS is 1. If the user sets this value to 0, then it disables Time Travel for the specified object. This parameter can be used to change the data retention period for a Snowflake object.

MIN_DATA_RETENTION_TIME_IN_DAYS

This parameter can be set by users to define the minimum retention period for the account. Like the data retention period, this value is defined by the number of days and is used to retain historical data for Snowflake objects. Its value can be as follows:

- 0 or 1 for Standard edition

- 0 to 90 for Enterprise or higher

The default value of this parameter is 0.

Note   This parameter does not apply to transient, temporary, or external tables, materialized views, and streams.

Managing the Data Retention Period

It is important to understand that as soon as users change the data retention period for a table, the change is applied to all active data and data in Time Travel. This is another important concept to understand, as explained here:

- *Increasing the retention period*: Consider an example for a table with a 20-day data retention period. Assume you add 10 days to this retention period (an increase from 20 days to 30 days). As a result, the original table data will be retained for 30 days instead of 20 days (10 extra days of data will be retained).

- *Decreasing retention period*: Let's discuss another example where we decrease the retention period of a table from 20 days to 1 day. As a result, data from 2 to 20 days is moved to Fail-safe mode.

Note   For datasets modified after the retention period is reduced, an updated data retention period applies.

LIST DROPPED OBJECTS

Users can list any dropped objects using the SHOW command along with the HISTORY keyword. For example, the following command can be used to show all the tables (including dropped tables) in SAMPLE_DB.SAMPLE_ SCHEMA (if users have required privileges).

SHOW TABLES HISTORY IN SAMPLE_DB.SAMPLE_SCHEMA.

Managing Storage Fees

Storage fees are calculated for each 24-hour period once data is changed. Based on the type of objects and defined data retention period, Snowflake maintains the historical data. As a best practice, Snowflake maintains only minimal details required to restore modified objects.

Users should consider using a temporary/transient table to minimize storage costs. The following are a few points worth mentioning here:

- Once temporary tables are dropped, data cannot be recovered.

- Once the Time Travel data retention period completes, transient tables' historical data cannot be recovered.

- Tables that are considered to stay for long should be defined as permanent to ensure they have an associated Fail-safe period.

Fail-Safe

The retention period for recovering historical data for Fail-safe is 7 days, and it starts (automatically) once the Time Travel retention period ends. Consider this as an additional layer of security provided by Snowflake, which can be used once all other data recovery options are exhausted. However, only Snowflake employees can access this data, and users should raise a case for it. It is recommended that this option should be used for best-endeavor purposes only, e.g., data lost due to operational failures.

Note   There is no option to configure the Fail-safe period (7-day default value). Also, it can be defined only for permanent tables.

Important Considerations

For Time Travel and Fail-safe, the following are a few important points to understand:

- By default, Time Travel is enabled for an account. However, one way to override this is by setting the DATA_RETENTION_TIME_IN_DAYS value to 0 (days).

- Between DATA_RETENTION_TIME_IN_DAYS and MIN_DATA_RETENTION_TIME_IN_DAYS (account level), a higher value is considered.

- The retention period of the temporary table is 24 hours or the remainder of the session, whichever is shorter. Once the session of the temporary table completes, the table is purged.

Summary

In this chapter, we discussed Time Travel and Fail-safe, which are revolutionary options provided by Snowflake for continuous data protection. They enable you to go back in time and view historical data. Additionally, we discussed data storage considerations and how to configure the data retention period.

CHAPTER 12

Continuous Data Pipelines

For organizations to do effective analysis, it is important to move data from different sources into a single data store. There are multiple stages involved to support this data movement, including capturing data in its raw format in the first stage followed by applying various transformations to the data to be consumed by users.

In general, there are three ways to load data: batches, micro-batches, and near real time. Batch processing refers to loading large volumes of data files (same type) together. Like batch processing, micro-batches also load data in batches, but the data size is small, and the execution cycle is more frequent. Real-time processing is ensuring data is available in near real time (data is processed as soon as it arrives).

A data pipeline is a sequence of steps used to move data from the source to the target while transforming it along the way. A continuous data pipeline focuses on removing manual steps from the data pipeline and adding automation (wherever possible) to enable near real-time data processing for faster business decisions. In this chapter, we will discuss various features provided by Snowflake to enable a continuous data pipeline.

© Ruchi Soni 2023
R. Soni, *Snowflake SnowPro™ Advanced Architect Certification Companion*,
Certification Study Companion Series, https://doi.org/10.1007/978-1-4842-9262-4_12

Snowflake Support for Continuous Data Pipelines

Continuous data pipelines provide near real-time data access to enable decision-making. Snowflake data pipelines can be batch or continuous. Snowflake enables a continuous data pipeline using the following options:

- *Continuous data loading*: This includes using the Snowpipe/Snowflake connector for Kafka/third-party integration tools.

- *Change data tracking*: This involves using the "stream" object for tracking data changes.

- *Recurring tasks*: Tasks refer to Snowflake's built-in scheduler to run SQL queries/stored procedures, etc. This adds control to execute things in a user-defined order. Usually, they are combined with streams to create a continuous data pipeline.

Streams

A stream object enables change data capture by recording DML changes made to tables and metadata about each change record. It stores an offset for a point-in-time snapshot for the source table and uses version history for change data capture. The change details captured contain the same row structure plus additional columns describing the change. Streams can be created to query change data on the following objects:

- Tables

- Views, including secure views

- Directory tables

- External tables

> **Note** Streams capture the offset of the source table only (no data).

Stream Columns

When queried, a stream accesses and returns the historic data with the same source row structure plus the following columns:

- METADATA$ACTION: The values can be INSERT or DELETE to explain the type of operation on the source table.

- METADATA$ISUPDATE: This is used to identify an UPDATE. The value can be TRUE or FALSE.

- METADATA$ROW_ID: This metadata field gives a row ID for a record to track changes.

Let's now understand streams using a simple example. Consider a simple Employee table and create a stream on it, as depicted in Figure 12-1.

```
create or replace table EMPLOYEE
(First_Name string,
Last_Name string,
City string);
create or replace stream EMPLOYEE_STREAM on table EMPLOYEE;
```

Figure 12-1. *Creating a stream on a table*

151

Now initially EMPLOYEE_STREAM is empty since we still have not inserted any records in the table. This is depicted in Figure 12-2.

```
SELECT * FROM EMPLOYEE_STREAM;
```

Figure 12-2. *Initial contents of a stream*

Now let's insert two records in the table and then view the contents of the stream. This is depicted in Figure 12-3.

```
INSERT INTO EMPLOYEE VALUES
('JOE','BDN','ATLANTA'),
('JAMES','GRANT','LASVEGAS');
SELECT * FROM EMPLOYEE_STREAM;
```

Figure 12-3. *Contents of a stream after an insert operation on the table*

Since this is an INSERT operation, the METADATA$ISUPDATE column value is FALSE.

Note   Different user SQL scripts can read data from the same stream without impacting the offset. However, when a stream is used in a DML transaction, it advances the offset value.

Here is a summary (which is important to understand from an exam perspective):

- In the case of a fresh INSERT to a table, the stream will contain one record with METADATA$ACTION=INSERT and METADATA$ISUPDATE=FALSE.

- If a record in a table is updated, then the stream will contain two records corresponding to one record in the source table: the original source record with the value for columns METADATA$ACTION=DELETE and METADATA$ISUPDATE=TRUE and the new record (for insert) with METADATA$ACTION=INSERT and METADATA$ISUPDATE=TRUE.

Note   For streams on views, change tracking must be enabled explicitly for the view and for the underlying tables to add the hidden columns to these tables.

For a transaction, a stream returns records from the current position to the start time of the transaction. If the transaction commits, then the stream position will move to the start time for the transaction; otherwise, there is no change to its position.

Stream Types

There are three types of streams provided by Snowflake.

- *Standard streams*: This is used to record all DML changes to the object. This is supported for standard and directory tables or views. The examples we discussed earlier in this chapter are related to standard streams.

- *Append-only*: This tracks row inserts only (update and delete are not recorded). This is supported for standard and directory tables or views.

- *Insert-only*: This is supported for external tables only. It records row inserts (delete not recorded).

Stream Data Retention Period

In previous chapters, we discussed the data retention period of a table. Now since a stream stores an offset (a point-in-time snapshot for the table), if the offset of a stream is outside the table data retention period, a stream is considered stale. This means you cannot access historical data for the source table (including unconsumed change records), and you need to create a new stream if you want to track changes to the table. Hence, it is advisable to consume records from a stream within the defined data retention period of the source table.

There are a few things that Snowflake internally takes care of for a stream. For example, if a stream is not consumed for a table with a retention period less than 14 days, then Snowflake automatically extends the table retention period (temporary) up to the stream offset (maximum of 14 days). Once the table is consumed, this retention period is again updated to the initial value defined for the table. In general, for a table, if both parameters DATA_RETENTION_TIME_IN_DAYS and MAX_DATA_

EXTENSION_TIME_IN_DAYS are defined, then the retention period of the stream would be the highest of the two. This is again converted to the default defined value for a table (once a stream is consumed).

Note   Streams on directory tables or external tables have no data retention period.

DESCRIBE STREAM

This command describes the columns in a stream. The syntax is DESCRIBE STREAM <NAME>, where <NAME> is the name of the stream.

Figure 12-4 explains the command executed on the EMPLOYEE_STREAM we created earlier.

```
DESCRIBE STREAM EMPLOYEE_STREAM;
```

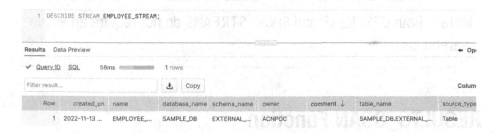

Figure 12-4. DESCRIBE STREAM command

In the output of the DESCRIBE STREAM command, the STALE column indicates whether the stream is currently expected to be stale, though the stream may not actually be stale yet.

Note   To query a stream, users will need an active warehouse, which will result in costs.

SHOW STREAMS

`SHOW STREAMS` is used to get the list of streams in the database, schema, and account (if you have required privileges). The output returns stream metadata and properties. Figure 12-5 explains this command executed on our account.

```
SHOW STREAMS;
```

Figure 12-5. *SHOW STREAMS command*

Note   Both `DESCRIBE` and `SHOW STREAMS` do not require an active warehouse.

RESULTS_SCAN Function

This function is not particularly associated with streams, but it is important to understand from an exam perspective. This function returns the result of a previous command like a table, so this can be used when users want to run SQL on the result of a command (like a table SQL). However, the previous command should be executed within 24 hours of this function execution.

In the function `RESULT_SCAN ( { '<query_id>' | LAST_QUERY_ID() } )`, the user can explicitly specify a query ID in the argument, or they can use the function `LAST_QUERY_ID` to get the ID for the last query executed.

Let's understand this using Figure 12-6.

```
CREATE OR REPLACE TABLE
SAMPLE_DATA AS
(select * from table (result_scan(last_query_id ())));
```

```
1   CREATE OR REPLACE TABLE
2   SAMPLE_DATA AS
3   (select * from table(result_scan(last_query_id())));
4
5   |
```

Results   Data Preview

✔ Query ID   SQL   1.63s ▬▬▬▬▬   1 rows

Filter result...   ⬇   Copy

Row   status

1   Table SAMPLE_DATA successfully created.

Figure 12-6. *Output of RESULT_SCAN function*

Tasks

In simple words, a *task* in Snowflake is more of a scheduler. It can run according to a set interval or a flexible schedule or manually and can execute SQL, stored procedures, and Java scripts, which can be combined with streams. Figure 12-7 depicts the task tree.

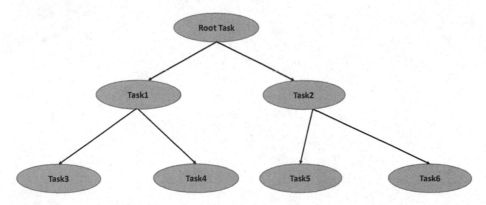

Figure 12-7. *Task tree*

> **Note**   A newly created task will be suspended by default. It is
> necessary to resume the task using the ALTER TASK <task_name>
> RESUME command.

Types of Tasks

Snowflake provides two types of tasks.

- *Serverless tasks*: As the name suggests, these tasks are
  managed by Snowflake. They use a virtual warehouse,
  and Snowflake automatically scales them up/down
  based on the complexity of the user query. To create
  serverless tasks, users should omit the WAREHOUSE
  parameter in the CREATE TASK command.

> **Note**   Serverless tasks cannot be used to invoke UDFs/stored
> procedures (SPs) written using Snowpark, which calls UDFs
> containing Java/Python code.

- *Tasks managed by users*: As the name suggests, these tasks are managed by users (use the WAREHOUSE parameter in the CREATE TASK command with the global EXECUTE MANAGED TASK privilege).

Note   ACCOUNTADMIN can provide privileges to the execute task.

Direct Acyclic Graph

A *direct acyclic graph* (DAG) refers to a unidirectional flow of tasks, starting with one root task that calls additional tasks. Except for the root task, each task in a DAG can have multiple predecessors and successors, and each task will start once the dependencies complete successfully. Each task should have the same owner, stored in the same DB and schema, and the EXECUTE TASK command can be run for manual task execution. It is also important to note that at a given time Snowflake executes only one instance of a task with a schedule.

Note   Along with the root task, a DAG can contain up to 1,000 tasks. A single task can have at most 100 predecessors and 100 successors.

SYSTEM$STREAM_HAS_DATA

This system function indicates whether the stream has data or not (contains CDC records). The following is the syntax used, where <S1> gives the stream name:

```
SYSTEM$STREAM_HAS_DATA('<S1>')
```

Users can combine streams and tasks together using this function. It can be used in the WHEN expression in the task definition. If the specified stream contains no change data, the task skips the current run, which can help save warehouse credits. For example, we can have a task running that checks the stream<sample_stream> for the presence of records to be processed using the system function SYSTEM$STREAM_HAS_DATA('<sample_stream>'). If the function returns FALSE, then there are no records to process, and the task will exit. If the function returns TRUE, the stream contains CDC records.

CREATE_TASK

This command creates/replaces a task. The following is the syntax used:

```
CREATE OR REPLACE TASK <task_name> [<Optional Parameters>]
AS <SQL>
```

Here <task_name> and <SQL> are mandatory parameters and there are various parameters associated with it. Some of the key optional parameters are as follows:

- WAREHOUSE provides the virtual warehouse for executing tasks (serverless).

- USER_TASK_MANAGED_INITIAL_WAREHOUSE_SIZE gives compute resource size for the first task run.

- SCHEDULE specifies a schedule for periodically running tasks.

Snowpipe

Consider Snowpipe as a virtual pipe between the source and the target. This is used for near real-time data loading (loading files as soon as they land on the stage). Internally, it uses the COPY statement for data loading, which identifies the source location of data files and target tables. One important point to consider is that it supports all data types (including semistructured data). There are two options for detecting staged files.

- *Cloud messaging*: This includes using a cloud event notifications service.

- *Calling REST endpoints*: This includes explicitly calling a REST endpoint with a Snowpipe name along with a list of data files. It requires key pair authentication using JSON Web Token (JWT).

CREATE_PIPE

This command is used to create a pipe. The following is the syntax for this command:

```
CREATE [OR REPLACE] PIPE <pipe_name>
  [AUTO_INGEST = VAL ]
  [ ERROR_INTEGRATION = INTEGRATION_NAME ]
  [ AWS_SNS_TOPIC = '<Val>']
  [ INTEGRATION = '<Val>' ]
  [ COMMENT = '<Val>' ]
  AS <copy >
```

Here <pipe_name> and <copy> are mandatory parameters, and the others are optional parameters. One important optional parameter is AUTO_INGEST, which can have a value of TRUE or FALSE where the TRUE value enables automatic data. If set to FALSE, then users should use Snowpipe REST APIs for data loading. We have already discussed this command in Chapter 4.

Note   Snowpipe supports data loading from internal and external stages. However, it does not support the user stage.

Summary

In this chapter, we discussed continuous data pipelines that focus on removing manual steps from the data pipeline and adding automation (wherever possible) to enable near real-time data processing. Snowflake enables continuous data pipelines using continuous data loading, change data tracking, and recurring tasks. We did a deep dive into the benefits and different options provided by Snowflake to enable this. This includes a deep dive into a few important concepts associated with streams, tasks, and pipes.

CHAPTER 13

Data Replication and Failovers

As we discussed in earlier chapters, data is an asset that has a major impact on the long-term success of an enterprise. Now as organizations become more and more dependent on data, it is important to introduce new ways to support high availability, backup, and disaster recovery for better data life-cycle management. There are two key performance indicators (KPIs) that are considered for disaster recovery: recovery time objective (RTO) and recovery point objective (RPO). RTO refers to the recovery time from a system outage, and RPO refers to the maximum amount of data loss acceptable to businesses. For example, assuming your system failed and the last backup was taken 12 hours earlier, then the RPO is 12 hours. Snowflake helps support these NFRs to minimize data loss using data replication.

Data replication involves creating a copy of the data across different databases on different servers to improve data availability and scalability. There are multiple benefits of data replication. If a system at one site goes down because of hardware issues or other problems, users can access data stored at another site. Data replication also allows faster access to data. Since data is stored in various locations, users can retrieve data from the nearest servers and benefit from reduced latency. It enables the accurate sharing of information so that all users have access to consistent data in real time. This chapter helps us understand how Snowflake manages data replication.

Data Replication

As mentioned earlier, the Snowflake data replication feature is used for data sharing across regions/clouds. This data is encrypted in transit between regions and cloud providers.

Users can enable Tri-Secret Secure for data replication. This uses a composite key, which is a combination of Snowflake and consumer-managed keys as an additional layer of security above the standard keys (Snowflake-managed and customer-managed). The idea here is to create a third (tri) level of security. This is explained using Figure 13-1.

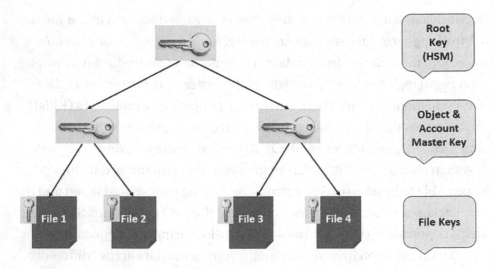

Figure 13-1. *Demonstrating Tri-Secret Secure*

Primary and Secondary Databases

In simple terms, the primary database is the one that contains user data, and the secondary database is a copy of a primary database that contains replicated data. In Snowflake, when you enable replication of a database, then it is considered a *primary database*, and the database where data is replicated is called a *secondary database*. Figure 13-2 explains the concept.

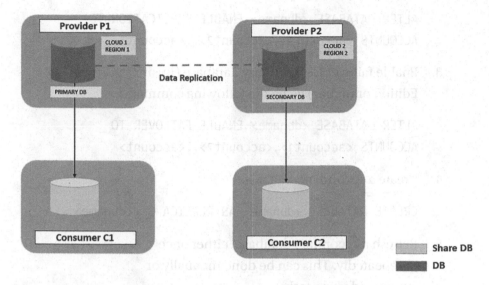

Figure 13-2. *Database replication*

Users can enable replication for permanent/transient databases. The database on which replication is enabled is called the *primary DB*, and the database where this data is replicated (across the region/cloud) is called the *secondary DB*. When you enable replication on a database, then a snapshot of its objects is copied to a secondary DB. Additionally, you need to create only one copy of the dataset per region irrespective of the number of consumers in the region.

Database Replication Steps

The following are the high-level steps to enable database replication. We have not included detailed instructions since those are not important for this exam:

1. Link your organization accounts.

2. Modify an existing database to act as the primary database using the following command:

```
ALTER DATABASE <dbname> ENABLE REPLICATION TO
ACCOUNTS <account1>,<account2>..<account>
```

3. Enable failover for a primary database (Business Critical Edition or higher) using the following command:

```
ALTER DATABASE <dbname> ENABLE FAILOVER TO
ACCOUNTS <account1>,<account2>..<account>
```

4. Create a secondary database.

```
CREATE DATABASE <dbname> AS REPLICA OF <dbname>
```

5. Refresh a secondary database, either once or repeatedly. This can be done manually or automated using tasks.

```
ALTER DATABASE <dbname> RFERESH;
```

You can execute the query DATABASE_REFRESH_PROGRESS to find out the status of replication for a specific database. This is explained later in the chapter.

Note   All data operations should be executed on the primary DB. The secondary database should be periodically refreshed from the primary DB snapshot.

Replication supports incremental refreshes; only changes that occurred during the last refresh are replicated, guaranteeing that replication operations are completed quickly. When users execute a replication command, a point-in-time snapshot of objects in the primary database is copied to the secondary database. Either users can use Snowflake serverless task to refresh secondary database at scheduled intervals or they can execute a manual refresh using the following command:

```
alter database <secondary_db_name> refresh;
```

The following are a few important points to know about data replication:

- If an external table exists in the primary database, replication is blocked.

- Stages and pipes cannot be replicated.

- Historical usage data of a primary database is not replicated.

- Databases created from shares cannot be replicated.

- Replicated streams can successfully track the change data for tables and views in the same database. However, the following source object types are not supported:

  - Directory tables

  - External tables

  - Tables or views in databases separate from the stream databases

  - Tables or views in a shared database

Dropping a Database

The following are a few important points to know:

- Use the DROP database command to drop the secondary database.

- If you want to drop the primary database, then promote the secondary database to a primary database and then drop the former.

Note   To drop a database, you should be the owner of it.

Important Considerations

There are a few important points to understand for data replication.

- Stored procedures and UDFs are replicated from a primary database to secondary databases.

- Snowflake performs automatic clustering if a secondary database with clustered tables is promoted as the primary database.

- Snowflake performs automatic materialized views maintenance in the primary database.

- Data replication fails in the following cases:

  - The primary database is Enterprise or a higher version, but any of the accounts for replication are lower editions.

  - Any primary database object has a dangling reference to a tag in another DB.

- To avoid unexpected behavior when running replicated tasks that reference streams, the user should create the tasks and streams in the same database, or if streams are stored in a different database from the tasks that reference them, include both databases in the same failover group.

- If a stream in the primary database is stale, then the replicated stream in a secondary database is also stale, and the user should re-create the stream in the primary database and again refresh the secondary database from the primary.

- After a primary database is failed over, if a stream in the database uses Time Travel to read a table version for the source object from a point in time before the last refresh timestamp, the replicated stream cannot be queried, and the changed data cannot be consumed.

- If the following conditions are met, a task is replicated to a secondary database in a resumed state:

  - When data replication begins, then the task is in a resumed state in the primary DB until replication completes. If a task is in a resumed state during only part of this period, it might still be replicated in a resumed state.

  - The parent database was replicated to the target account along with role objects in the same, or different, replication, or failover group.

- Time Travel and Fail-safe data are maintained independently for a secondary database and are not replicated from the primary database.

- Cloned objects are replicated physically to the secondary database.

- Database replication is supported for the database, and other object types can be replicated using account replication.

- For database replication, the billing amount of data changed in the primary database and the frequency of secondary database refresh are considered.

- Time Travel and Fail-safe data are maintained independently for a secondary database and are not replicated from the primary database.

Note   A database refresh operation can require several hours or longer to complete depending on the amount of data to replicate.

Replication Functions

The following are a few important functions that can be used during database replication:

- REPLICATION_USAGE_HISTORY:

  This function is used to get the replication usage history of a database. The following is the syntax used:

  REPLICATION_USAGE_HISTORY('start_date','end_date','database')

  All three parameters are optional. If they are specified, then it gives the usage history of the database within the start_date and end_date. The result of executing this function is the database name, number of credits consumed by replication activity, and number of bytes replicated.

  To run this function, you need the ACCOUNTADMIN role or MONITOR USAGE privilege.

> **Note**   If you don't specify any parameter, then this function returns the last 14 days of replication usage activity for a user account.

- DATABASE_REFRESH_PROGRESS

  As mentioned earlier, this function can be used to get the replication status of a database. This function can be called with the name of the secondary database to know its replication status, as shown here:

  DATABASE_REFRESH_PROGRESS('db_name')

  Here db_name is the name of the secondary database to get the replication progress details. This function returns a JSON structure with details of the data replication status.

Database Failover

Using database failover, customers can fail over databases to an available region or cloud provider to continue business operations in the event of a massive outage that disrupts the cloud services in a region or for a specific cloud provider. Snowflake database replication and Snowflake database failover services occur in real time, and the time to recover doesn't depend on the amount of data. If a disaster occurs in one region or on one cloud service, businesses can immediately access data that is already replicated in a different region or cloud service.

Snowflake also provides client redirect functionality that allows redirecting client connections to user accounts in different regions for disaster recovery. This requires a Snowflake connection object that stores the "connection URL" to establish the required connectivity. The following command can be used to enable failover for a database:

```
ALTER DATABASE ENABLE FAILOVER TO ACCOUNTS <account1>
<account2>...<accountn>
```

Note   This requires a Business-Critical or higher Snowflake edition. Only users with the ACCOUNTADMIN role can enable and manage failover for a database.

Summary

Snowflake database replication and failover features provide instant fallback operations for immediate data recovery. If an outage occurs, customers can initiate a database failover to elevate secondary databases in the available region to serve as primary databases for workloads. Once the outage is resolved, customers can return to normal business operations. This provides you with a consistent and reliable solution to manage data, ensuring high availability and scalability.

CHAPTER 14

Managing Accounts and Security

As data becomes more and more important for the organization and with the sharp rise in cybercrime, data security is a major concern for every enterprise. Now Snowflake believes in continuous data protection (CDP) by providing a set of features that help protect data stored in Snowflake at every stage within the data life cycle. Another important aspect to understand is user and account management. Snowflake supports role-based access control, which forbids any privilege assignment directly to users. Instead, here privileges are assigned to roles that are assigned to users.

In this chapter, we will discuss the activities associated with managing your account in Snowflake and the different aspects of data security and how Snowflake supports it.

Account Management

Snowflake creates a hierarchical structure for account management starting with Organization followed by Account and others. Let's understand these elements and how are they structured in detail.

© Ruchi Soni 2023
R. Soni, *Snowflake SnowPro™ Advanced Architect Certification Companion*,
Certification Study Companion Series, https://doi.org/10.1007/978-1-4842-9262-4_14

- *Organization*: This is a first-class Snowflake object that links the accounts owned by users' business entities. It enables users to get a central view of all accounts. The ORGADMIN system role is responsible for managing operations at the organizational level. Users can use the organization_usage schema to create a consolidated view of compute, storage, and data transfer usage across all accounts within the organization. They can even drill into each account using the account_name column in each view. This enables them to monitor account usage and spend from all accounts in a single view. You can add new accounts in your organization using the Create New Account option on the Organization page.

- *Account*: An account is created within an organization using the CREATE ACCOUNT command. When creating an account, users can specify a cloud platform, a region, and a Snowflake edition. There can be multiple reasons for an architect to create separate Snowflake accounts within an organization. For example, a separate account can be created to test features that are in private preview for the detection of errors, to create separate high-tiered accounts for cost management, etc.

The region determines where the data in the account is stored and where the compute resources used by the account are provisioned. Each account in an organization has a specific Snowflake edition that determines its available features and level of service. The following are a few important aspects to understand for efficient account management:

- An account identifier uniquely identifies a Snowflake account throughout the global network of Snowflake-supported cloud platforms and cloud regions.

- Account names must be unique within one organization.

- Three types of parameters can be associated with an account.

    - Account parameters that affect the entire account

    - Session parameters that default to users and their sessions

    - Object parameters that default to objects

Note   By default, the maximum number of accounts in an organization cannot exceed 25.

Each account in an organization can have its own set of users, roles, databases, and warehouses.

Each database has multiple schemas.

Each schema contains different objects such as tables, views, stages, etc.

Figure 14-1 gives a pictorial representation of this.

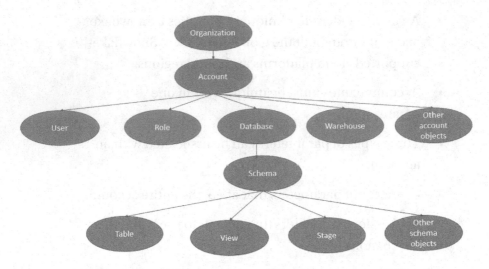

Figure 14-1. *Account management hierarchy*

User Management

Snowflake uses role-based access control for creating and managing users. A few important user roles are mentioned here:

- The USERADMIN system role is required to create users.

- To modify the properties of a user, you need a role with OWNERSHIP and global CREATE USER privileges.

- By default, every Snowflake user has a PUBLIC role, and based on specific access requirements, they can be assigned additional roles while you are creating a user or later.

- As a best practice, the ACCOUNTADMIN role should be assigned to at least two users. However, this should not be the default role and should be properly managed and controlled.

- It is advisable to enable multifactor authentication (MFA) for better security. Here, in addition to Snowflake credentials, users should add a passcode from the Duo mobile application.

- By default, Snowflake locks a user account for 15 minutes in the case of five consecutive wrong login attempts.

Important Commands

Here are some important commands:

- CREATE USER: As the name suggests, this command is used to create a new user. To execute this command, you need the USERADMIN role or the account's CREATE USER privilege.

- DESC USER: This command describes a user, including the current values for the user's properties, as well as the default values. Figure 14-2 explains the output of this command.

  DESC USER RUSONI;

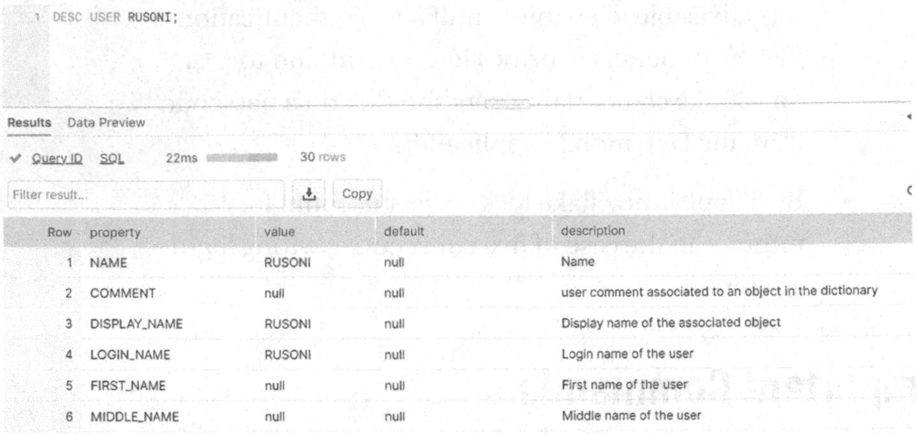

```
1  DESC USER RUSONI;
```

Results   Data Preview

✓ Query ID   SQL   22ms   30 rows

Filter result...   ⬇ Copy

Row	property	value	default	description
1	NAME	RUSONI	null	Name
2	COMMENT	null	null	user comment associated to an object in the dictionary
3	DISPLAY_NAME	RUSONI	null	Display name of the associated object
4	LOGIN_NAME	RUSONI	null	Login name of the user
5	FIRST_NAME	null	null	First name of the user
6	MIDDLE_NAME	null	null	Middle name of the user

Figure 14-2. *DESC USER*

To execute this command, you need the user's OWNERSHIP privilege. Individual users can see their own properties by executing this command with their login name.

- SHOW USERS: As the name suggests, this command lists all the users in the system. It can be executed only by users with a role that has the MANAGE GRANTS global privilege. Additionally, this command does not require a running warehouse to execute.

- ALTER USER: This command is used to alter properties for an existing user. To execute this command, you need the user's OWNERSHIP privilege.

- DROP USER: This command removes the specified user from the system. These dropped users cannot be recovered and must be re-created.

Access Control Framework

There are two types of access control models.

- *Discretionary access control (DAC)*: Here access is granted directly to an object by its owner.

- *Role-based access control (RBAC)*: Here privileges are assigned to roles that are assigned to users. Users can access objects through these defined roles only (no direct access).

Snowflake combines aspects from both RBAC and DAC models. To understand the Snowflake access control framework, first let's understand the elements.

Securable Objects

Users are granted access to securable objects through RBAC (privileges assigned to roles assigned to users). By default, these objects are owned by roles used to create them, and these roles are allocated to other users for access. In a regular schema, the owner role has all the privileges on the object, and in a managed access schema, the schema owner, or a role with the MANAGE GRANTS privilege, can grant privileges on schema objects.

RBAC in Snowflake supports role inheritance for users. Let's consider an example where Role 1 is granted to Role 2 and Customer A is granted to Role 2. Now in every session, Customer A can use the sum of privileges granted to Role 1, Role 2, and the PUBLIC role, and all of them are effective during a session. This is called *role inheritance*. Privilege inheritance is possible only within a role hierarchy.

Let's consider an example of a role ETL, which has full access (read/write/create) to a database object. Let's assume we want to assign this role to a user developer. The following is the steps that can be used:

Create role ETL;

Grant All on database sample to role ETL;

Grant role ETL to user developer;

Below command can be used to revoke this role

Revoke role ETL from user developer;

Roles

As per the standard RBAC model, privileges on objects are granted to roles that are assigned to users or other roles (hierarchy). These roles can be system-defined or user-defined (customer roles). You cannot drop or revoke privileges for system-defined roles as these are created for account management. The different system-defined roles supported by Snowflake are mentioned here:

- ORGADMIN: This manages operations on the first-class Snowflake object (organization). This is used to create and view account and usage information.

- ACCOUNTADMIN: This is the most powerful role in the hierarchy for account-level parameter configuration (including billing). It encapsulates both SYSADMIN and SECURITYADMIN roles.

Note   As a best practice, this role should be assigned to a limited number of users in an organization. The general recommendation is for two users with the ACCOUNTADMIN role to ensure the account always has at least one user who can perform account-level tasks (another user is a backup).

- SECURITYADMIN: This can be used to create and monitor users and roles and manage grants.

- USERADMIN: This includes privileges for the user and role management that includes creating and managing users and roles.

Note   By default, the first user (admin) is assigned the ACCOUNTADMIN role when the account is created. All other users should be created with USERADMIN or a role with global CREATE USER privileges.

- SYSADMIN: This is used to create warehouses, databases, and other database objects.

- PUBLIC: This is the default role to own securable objects.

Figure 14-3 explains the roles hierarchy, which is important to understand from an exam perspective.

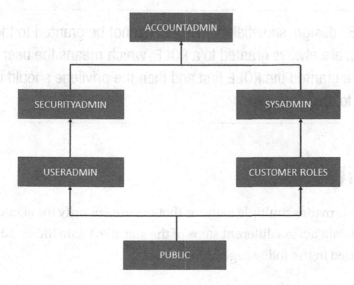

Figure 14-3.   *Role hierarchy*

181

Custom roles can be created using a USERADMIN (or higher) role or a role to which the CREATE ROLE privilege is granted. Additionally, they should be created based on the principle of "least privilege" in accordance with departments in your organization to provide required access only. They can be used by architects to extend these out-of-the-box role hierarchies provided by Snowflake.

Privileges

Privileges are used to control access for an object. Users are granted access to securable objects through privileges assigned to roles that are assigned to users.

Users

This refers to the individual user who accesses the object. For enhanced data security, Snowflake does not recognize the "superuser" role that can bypass any access checks.

Note   By design, snowflake privileges cannot be granted to the USER but are always granted to a ROLE, which means the user should be granted the ROLE first and then the privilege should be granted to the ROLE.

Security

Snowflake provides multiple features that ensure security for accounts, users, and data across different steps of the standard data life cycle. These are explained in the following section.

Column-Level Security

Column-level security includes applying a masking policy to a field that includes both dynamic data masking and external tokenization. Dynamic data masking is used to mask table data while executing user queries, so users are unable to see data for specific columns in plain text. This helps with data authorization and governance and provides an option to selectively protect the contents of a column for display; e.g., organizations' personal identifiable information (PII) data should be accessible only to the HR business unit and masked from all other departments. This can be done by creating a masking policy and applying it on the PII fields, enabling it for everyone except individuals with Department=HR.

Tokenization, on the other hand, provides selective data protection using undecipherable tokens. It allows data tokenization before data loading and detokenization during query runtime. From an exam perspective, you should know the key difference between the two, and a practical implementation is not part of the exam. Hence, we have not discussed this in detail.

Row-Level Security

Row-level security (RLS) is supported by Snowflake using row-access policies. Like column-level security that is applied on fields, RLS focuses on the rows. It is a schema-level object that determines whether a given row in a table/view is accessible to users. This is implemented using row-access policies that contain an expression to check which rows should be visible to users. For example, consider a geography table that contains records for various geographies; say we want to restrict access to records based on geography. This can be done by applying a row-access policy on the table to provide the required level of access.

Object Tagging

In simple terms, a *tag* is an object that can be assigned to another
object. Tags enable data stewards to track sensitive data through either
a centralized or decentralized data governance management approach.
Once you define a tag and assign it to Snowflake objects (tables, views,
columns, warehouse, etc.), the objects can be queried to track usage on the
objects to facilitate auditing. When assigned to a warehouse, tags enable
accurate resource usage tracking. Let's understand this with an example.
Figure 14-4 creates a tag that lists the environments.

```
create or replace tag environment
allowed_values 'DEV', 'QA','PROD';
```

Figure 14-4. *Creating a tag*

Now let's assign this tag to a warehouse, as shown in Figure 14-5.

```
alter warehouse COMPUTE_WH set tag environment = 'DEV';
```

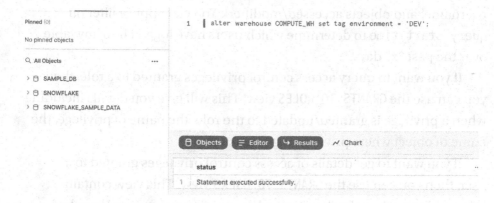

Figure 14-5. *Assigning a tag to a warehouse*

Data Classification

As the name suggests, data classification analyzes and categorizes information stored in the columns in database tables and views. Typical use cases of data classification include marking PII/sensitive data, configuring security controls, setting masking policies, anonymizing personal data, etc. The data classification process requires compute resources. Data can be classified into the semantic category (identifies a column as storing personal attributes) and the privacy category (sensitive personal data). We are not going into the details of this since that is not important from an exam perspective.

Access History

This includes querying the account usage ACCESS_HISTORY view to get the access history of the user. It contains a single record per SQL statement, which explains the fields accessed by the SQL. These records can be used to understand a user data access pattern, which includes frequently accessed data that can be used for compliance and auditing purposes. This view contains information related to a query ID, query start time,

username, and objects accessed/modified. You can apply a filter on
query_start_time to determine which users have logged into any table
over the past "x" days.

If you want to query access control privileges granted to a role, then
you can use the GRANTS_TO_ROLES view. This will give you details including
when a privilege is granted/updated to the role, the name of privilege, the
name of object where privilege is granted, etc.

If you want to get details of access control privileges granted to a
user, then you can use the GRANTS_TO_USERS view. This view contains
information such as the data/time when the role was granted/revoked,
the name of the role granted, the name of the user whom the privilege is
granted, etc.

It is worthwhile to mention two important schemas, as
mentioned here:

- *Information_Schema*: This is a read-only schema and
  is created by Snowflake for every database in the user
  account. It contains views for database and account-
  level objects and contains metadata for those objects.
  This schema also contains table functions to get usage
  information (account level) and historical details of
  Snowflake objects.

- *Account Usage*: This is also a read-only schema
  that contains views for the metadata of objects and
  accounts usage metrics.

Note   The information schema does not include objects that are
dropped. On the other hand, account usage views include details
of dropped objects. Also, the former includes 7 days to 6 months of
historical data (Varies by views/table) and later includes 1 year of
historical data.

Data Security

Snowflake automatically encrypts data ingested in Snowflake tables.
It is encrypted using AES-256 encryption, which includes encrypting
all files stored in internal stages. Additionally, in Business-Critical and
above editions, Snowflake also supports encrypting data using customer-
managed keys.

Network Policy

A network policy can be used to manage network configurations to the
Snowflake service and control user access based on IP address. A network
administrator can create a network policy to enable or restrict access
to a list of IP addresses. Once created, these network policies should be
activated at the account or individual user level. Please note that currently
only Internet Protocol version 4 is supported by Snowflake.

SSO

Single sign-on (SSO) enables users with a cloud account to seamlessly
access services provided by different partners, without any need for a
separate login for an individual partner. Snowflake supports federated
authentication using SSO, enabling users to authenticate through an
external, SAML 2.0–compliant vendor. Okta and Microsoft ADFS provide
native Snowflake support for SSO.

Summary

Snowflake is a secure, cloud-based data warehouse. In this chapter, we
discussed various features provided by Snowflake for continuous data
protection (CDP) that includes support for user and account management.

This also includes an understanding of the hierarchical structure for account management starting with the customer organization followed by accounts. We also discussed both access control frameworks (discretionary access control and role-based access control) and a few aspects of managing security.

CHAPTER 15

Query Profiles and Tuning

Organizations need high-speed access to real-time data for strategic business decisions as faster decision-making drives high business value. The standard way for users to access data is by running queries against the data store. Query optimization is important to access databases in an efficient manner. This refers to the process of selecting an effective way to execute the query.

This chapter helps us to understand how users can analyze queries using a query profile, which is a tool provided by Snowflake to analyze performance.

Query Profiles

A *query profile* is a useful option provided by Snowflake to understand the details of an executed query; it enables users to understand query behavior. It is important to understand from an exam perspective since you will get a few scenario-based questions on the exam where you will be asked to identify a query bottleneck based on the information provided in a query profile.

© Ruchi Soni 2023
R. Soni, *Snowflake SnowPro™ Advanced Architect Certification Companion,*
Certification Study Companion Series, https://doi.org/10.1007/978-1-4842-9262-4_15

The following are the steps you can use to access a query profile:

1. Open the History tab, as explained in Figure 15-1.

Figure 15-1. *History tab*

2. Now let's click the query ID of the following query on the History tab, as demonstrated in Figure 15-2:

```
Create or replace table
sample_data as
(select * from table(result_scan(last_query_id())));
```

Status	Success
User	RUSONI
Warehouse	TEST_DW
Start Time	11/13/2022, 4:02:29 PM
End Time	11/13/2022, 4:02:30 PM
Total Duration	1.6s
Scanned Bytes	0
Rows	1
Query ID	01a84798-0605-2787-0000-0ed50454a236
Session ID	16308038438702

SQL Text

```
1  create or replace table
2  sample_data as
3  (select * from table(result_scan(last_query_id())));
```

Query Result

The results for this query have expired.

Figure 15-2. *Viewing the details of a query*

3. Now click the Profile tab to get the query profile details, as explained in Figure 15-3.

Figure 15-3. *Query profile tab*

The query profile consists of the following key elements:

- *Steps*: Generally any user query involves a series of execution steps/paths. This query profile allows you to move across different steps and view the details of each individual step/path in a separate panel.

- *Operator Tree*: This refers to the middle section that allows users to view a diagrammatic representation of the nodes along with the relationship between nodes. This includes a collapsible panel that lists nodes by execution time in descending order.

- *Node List*: This section contains a list of operator nodes with their execution time.

- *Profile Overview*: When you select a node in the operator tree, then this section will provide details of the specific component that includes the following information:

  - *Node Execution Time*: This gives details about any tasks that consumed query time.

  - *Statistics*: This provides details about query statistics.

191

- *Attributes*: This provides component-specific information, e.g., full table scan, join type, etc.

Figure 15-4 shows the profile overview pane that opens when you click the CreateTableAsSelect node.

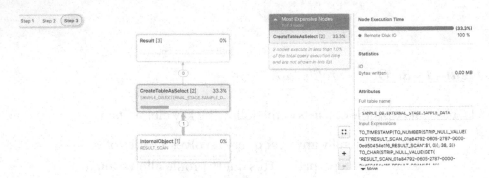

Figure 15-4. *Profile overview pane*

Profiling Information

The detail panel provides the following profiling information that can be used to analyze the performance of queries:

- Query execution time

- Detailed statistics

It is important to understand these to interpret queries, identify bottlenecks, and give recommendations for better query performance. It is also important from an exam perspective to answer scenario-based questions and give options for how to cut down query processing time. Although there is a big list of parameters to analyze query performance, a few of them are worth mentioning.

- *Percentage scanned from cache*: This refers to how much data is accessed from the local disk cache. If in a query a major portion of the data is retrieved from the local disk cache instead of from the remote disk, it means faster query performance. This is a technique that users can use when designing ETL queries as co-locating queries, which retrieve data from the same tables that improve the query performance.

- *Remote disk I/O*: This refers to the time spent on remote I/O. In principle, the smaller the value for accessing the remote disk, the faster the query. When the query is executed for the first time, some of the data is cached in the local SSD disk of the virtual warehouse, and when the user executes the query again, a major portion of the data is retrieved from the local disk cache instead of from the remote disk.

- *Bytes spilled to remote storage*: This refers to data that spills over to remote storage. If your virtual warehouse is not sized properly and does not have enough memory, the intermediate results start spilling to the remote disk, which impacts query performance since the query is retrieving results from the remote disk instead of memory.

- *Partitions scanned and partitions total*: This refers to the total partitions scanned from a total number of table partitions (partition total). These two parameters help us understand the efficiency of pruning. If the partitions scanned are a small fraction of the partitions total, then the pruning is efficient.

The following is an exhaustive list of profiling information.

Execution Time

- *Processing*: This refers to the data processing time.

- *Local Disk IO*: This refers to the time spent on local disk access.

- *Remote Disk IO*: This refers to the time spent on remote disk I/O.

- *Network Communication*: This refers to the time spent on network data transfer.

Statistics

This includes the statistics associated with I/O, DML, pruning, etc. The following are the detailed options:

- *Scan progress*: Percentage of data scanned

- *Bytes scanned*: Total number of bytes scanned

- *Percentage scanned from cache*: Data percentage scanned from the local disk cache

- *Bytes written*: Total bytes written

- *Bytes written to result*: Total bytes written to result object

- *External bytes scanned*: Bytes read from an external object

- *Partitions scanned*: Number of partitions scanned from table

- *Partitions total*: Actual number of partitions in a table

- *Bytes spilled to local storage*: Data spilled (volume) to local storage

- *Bytes spilled to remote storage*: Data spilled (volume) to remote storage

- *Bytes sent over the network*: Data sent over the network

- *Number of rows inserted*: Count of rows inserted into a table

- *Number of rows updated*: Count of rows updated in a table

- *Number of rows deleted*: Count of rows deleted from a table

- *Number of rows unloaded*: Count of rows unloaded during data export

Summary

In this chapter, we discussed how to access a query profile and did a deep dive across various sections of the query profile, including the steps, operator tree, node list, and profile overview. We also discussed how to interpret the profiling information for query performance and how to identify bottlenecks, and we gave recommendations for faster query execution.

CHAPTER 16

Performance and Resource Optimization

In this chapter, we will discuss a few performance and resource optimization techniques provided by Snowflake. This includes the Snowflake Search Optimization service, which improves the performance of certain types of lookup and analytical queries. We'll also dive deeper into Snowflake caching.

Snowflake Search Optimization

The Search Optimization service improves the performance of specific types of queries, as mentioned here:

- Point lookup queries that return a few distinct rows. It allows faster response for critical dashboard/reports, which uses point lookup queries with selective filters on high-volume tables.

- Substring and regular expression searches.

© Ruchi Soni 2023
R. Soni, *Snowflake SnowPro™ Advanced Architect Certification Companion*,
Certification Study Companion Series, https://doi.org/10.1007/978-1-4842-9262-4_16

- Queries on fields in VARIANT, OBJECT, and ARRAY columns that use certain types of predicates. When VARIANT support for the Search Optimization service is configured for columns in a table, the Search Optimization service automatically includes VARIANT, OBJECT, and ARRAY columns in a search access path.

- Queries that use selected geospatial functions.

In addition, the Search Optimization service does the following:

- Search optimization improves the performance of queries, which typically run for a few seconds or longer.

- Search optimization works best for a query in which at least one of the columns used in the query filter operation contains a minimum of 100 GB distinct values.

- It does not provide many query performance benefits for tables smaller than 10 GB.

The Search Optimization service runs in the background to collect metadata information about the columns and populates search access paths that include data needed to be used to perform the table lookups for the query. These search access paths are updated automatically when the underlying table data changes.

Note   Search optimization is best suited for cases where users query tables on columns beside the cluster key.

Access Control for the Search Optimization Service

The following privileges are required to implement (add/remove) search optimization for a table:

- Table OWNERSHIP privilege

- Schema ADD SEARCH OPTIMIZATION privilege

SELECT privileges are required on the specific table to use it for a query.

Optimization Techniques

We do have other query optimization techniques that include clustering, materialized views, and search optimization. Now let's understand when to use what.

- Clustering a table can improve the performance of range searches and equality searches if they are on the clustering key. However, search optimization speeds up equality search only.

- Clustering can be done on a set of keys, but search optimization is enabled for all columns.

- Clustering adds additional compute costs to run the background process for reclustering, whereas search optimization adds both compute and storage costs.

- Materialized views can improve the performance of both equality searches and range searches.

Figure 16-1 provides additional details (storage and compute) for the three types of optimization techniques.

Optimization Technique	Storage Cost	Compute Cost
Search Optimization	Yes	Yes
Clustering	No	Yes
Materialized Views	Yes	Yes

Figure 16-1. *Optimization techniques*

Note   The Search Optimization service is available for Snowflake Enterprise and above editions.

The Cost of Search Optimization

Both storage and compute costs are required for the Search Optimization service, as mentioned here:

- Since search optimization creates a search access path data structure, it requires storage for each table where it is enabled.

- Adding and maintaining search optimization to a table consumes resources.

SYSTEM$ESTIMATE_SEARCH_ OPTIMIZATION_COSTS

This function returns the costs required to enable and configure specific columns for search optimization on a table.

The following is the syntax used:

```
SYSTEM$ESTIMATE_SEARCH_OPTIMIZATION_COSTS('<t1>' , <search_
method_with_target>)
```

Here, t1 is the table name to estimate the search optimization cost, and <search_method_with_target> specifies a search method and target for a column configuration.

Search Optimization Limitations

The Search Optimization service is not supported for the following objects:

- External tables
- Materialized views
- Column concatenation
- Analytical expressions
- Casts on table columns

Note   As a best practice, you should first add search optimization on a few tables and then expand it based on performance and cost benefits achieved.

Caching

Caching as a concept was already discussed in Chapter 2 of this book. However, since it is one of the ways in which Snowflake supports optimization, it's worth mentioning again. On the exam, you will get a few scenario-based questions about different types of caching, so you should understand which cache to use when.

Snowflake provides three layers of caching, as explained here:

- *Result cache*: When a query is executed, the result is
  cached (in the result cache) for a period, after which
  they are purged from the system. If a user repeats a
  query that has already been run and the data in the
  table has not changed since the last time that the query
  was run, then this result cache is assessed for the query.
  Query results are cached in the result cache for 24
  hours only, after which the cache should be rebuilt for
  the same query.

  Result caches are used by all the virtual warehouses,
  which ensures users can share the same query results
  when the underlying data has not changed. It is
  important to mention that each time the persisted result
  is reused, Snowflake resets the 24-hour retention period
  up to a maximum of 31 days from the date and time
  that the query was first executed after which the results
  are purged.

Note   By default, result reuse is enabled. To disable this, use the
USE_CACHED_RESULT session parameter (`alter session set
USE_CACHED_RESULT = FALSE`).

- *Local disk cache, warehouse cache, SSD cache, and
  data cache*: A virtual warehouse is a cluster of compute
  instances with local SSD. Now the data cache is the SSD
  cache that is part of the nodes of the virtual warehouse
  and is used to cache data used by SQL queries. When
  a new query is executed (result cache cannot be used),

the data needed for the query is retrieved from the remote disk storage and cached in SSD and memory. When users execute a query for the first time, the results of the query are retrieved from the remote storage, and part of the results are cached in the local SSD storage of the compute instance. When users execute the query again, part of the results are retrieved from this data cache.

Since it is associated with virtual warehouses, its size depends on the warehouse size (the bigger the warehouse, the larger the cache, and vice versa).

Note   When a virtual warehouse is suspended, it loses the data cache, which is rebuilt when the warehouse is resumed.

- *Remote disk cache*: This cache refers to long-term data storage. It is responsible for data resilience. Snowflake organizes data as compressed micro-partitions optimized for storage, which are adjoining storage units. The micro-partition size can be between 50 MB and 500 MB (uncompressed). Data is organized in a columnar fashion with Snowflake storing related metadata.

Figure 16-2 explains the different caching layers provided in Snowflake.

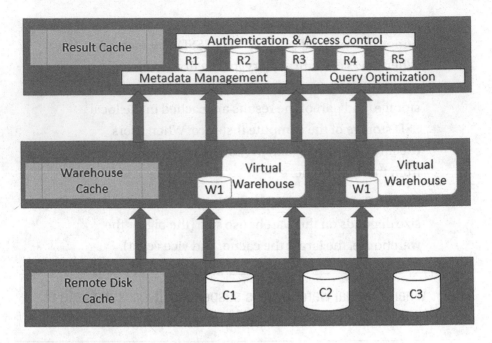

Figure 16-2. *Three layers of caching*

Other Techniques

Here are some techniques:

- As a software-as-a-service platform, Snowflake takes care of resource optimization and performance tuning and eliminates basic maintenance tasks, which reduces IT administrator overhead.

- Users can create usage monitoring queries to identify the warehouses, users, and queries that are consuming more credits than expected to take the necessary steps to reduce their consumption.

- Snowflake's resource monitoring service helps enable billing transparency for customers.

- Snowflake also provides alerting and usage management across the user, resource, workload, and account levels.

Summary

In this chapter, we discussed the performance and resource optimization techniques provided by Snowflake. This includes the various optimization techniques, including the Search Optimization service, how caching improves performance, and the additional options provided by Snowflake for resource optimization and query tuning.

Summary

CHAPTER 17

Snowflake Best Practices

Snowflake is a comprehensive software-as-a-service (SaaS) data platform. The beauty of the Snowflake architecture is a hybrid of the shared-disk and shared-nothing models. Over the years I have supported multiple Snowflake assignments, which include migration as well as modernization work.

In every client discussion, one important thing that gets discussed is how can you optimize our workloads. In simple terms, optimization is the art of finding a balance between what your customers want and what your business can afford within the available constraints of the system. Based on our detailed discussion in previous chapters, in this chapter, we will discuss different optimization options provided by Snowflake.

Best Practices for Query Optimization

In simple terms, *query optimization* is the process of choosing the most efficient way to execute a query. Snowflake provides different features, which if used correctly can significantly improve the performance of queries. Let's discuss a few options for query optimization.

© Ruchi Soni 2023
R. Soni, *Snowflake SnowPro™ Advanced Architect Certification Companion*,
Certification Study Companion Series, https://doi.org/10.1007/978-1-4842-9262-4_17

- To determine the right size of the warehouse for a specific query need, you should start with the smallest warehouse and adjust as per the optimized use and cost. Avoid focusing on the warehouse size as Snowflake utilizes per-second billing, so users can run larger warehouses and simply suspend them when not in use.

- Scale up using a larger warehouse size for complex queries, and scale out for concurrency using the Snowflake multicluster feature.

- Group and execute similar queries on the same virtual warehouse to maximize query data cache reuse.

- Use Snowflake's Auto Suspend and Auto Resume features. Warehouses can be set to automatically suspend when there's no activity after a specified period.

- Do not use the max-out limit for a multicluster warehouse.

- Leverage the dimension columns most frequently used in the filter as cluster keys and additional columns used in joins.

- Leverage materialized views for frequently used complex queries and for a fixed set of data.

- Leverage clones to implement fixes and then utilize SWAP functionality.

- Use search optimization for tables with low throughput and align optimization on columns with low cardinality.

- A materialized view can improve the performance of equality and range searches (for the subset of rows and columns in the view).

- Avoid using `ORDER BY` in subqueries as it does not impact the results and slows down performance. You should use `ORDER BY` in a top-level `SELECT` only.

- The `COPY` command is more performant than an `INSERT` command, as `COPY` uses bulk and `INSERT` is row by row.

Best Practices for Data Loading

As discussed in previous chapters, Snowflake provides two options for data loading: bulk versus continuous. As the name suggests, bulk loading is used to load bulk data (in batches) from Snowflake's internal/external stage to Snowflake tables, whereas continuous loading is used to load small volumes of data and incrementally make them available for analysis.

- Snowflake provides different table types like permanent, transient, and temporary. Always use a table type that best suits the requirements; for example, use temporary or transient tables for storing data that does not need to be maintained for extended periods of time.

- To prevent any unexpected storage charges, explicitly drop temporary tables once they are no longer needed. Users can also explicitly exit the session in which the table was created to ensure no additional charges.

- A pipe does not support the PURGE copy option, which means Snowpipe cannot delete staged files automatically when the data is successfully loaded into tables. To remove staged files, users should periodically execute the REMOVE command to delete the files.

- Utilize the STRIP_OUTER_ARRAY function to align the row size in VARIANT columns.

- Use Auto Ingest in Snowpipe for continuous loading. Also, consider staging the files once per minute to achieve cost savings and optimal latency in Snowpipe.

- Always prefer the Date and Numeric columns for deleting and inserting.

- Split structured data files into smaller chunk sizes of 100 to 250 MB compressed for parallel processing. Keeping files below a few gigabytes is better to simplify error handling.

- Unless the user explicitly specifies FORCE = TRUE, the COPY command ignores files that are in the staging area already loaded into the table.

- Staged files can be deleted from the stage using the REMOVE command. This helps improve the performance during data loading since it reduces the number of files for the COPY command to scan.

- For small tables less than 1 TB, there is no need to define a clustering key since on such a small table, natural clustering works great.

Best Practices for Cost Optimization

Here are some best practices:

- *Natural pre-sorting of data*: Sorting data before you load it into Snowflake enables Snowflake to automatically partition based on natural ingestion order. Pre-sorting of data load helps you in performance and helps you avoid cluster keys on a table, reducing the underlying cost. Pre-sorting of data may add overhead on ETL, but for the historical load, it can be done in batches.

- *Auto-clustering*: Start with a limited number of tables with clustering since the cluster refresh activity adds up to your storage and compute cost. Cluster tables only when you see a lag in performance and clustering tables would add benefit in the scenario. Snowflake recommends clustering the table from the lowest cardinality to the highest cardinality column (not more than three columns).

- *Materialize view*: MV adds up to your storage and computes (MV refresh activity) cost. So, you need to measure MV maintenance cost benefits over the performance cost.

- *Search optimization*: Since the search optimization service cost you storage and compute for maintenance, you need to decide on scenarios when this service is required on tables.

- *Database replication*: Database replication is an important service required for disaster recovery and other purposes. Using this service incurs extra charges for storage, computing, and data transfer activity. Data transfer between different regions incurs transfer charges, and transfer rates are different for the same or different continents. You may separate your intermediate staging or temporary tables into separate staging databases, thus avoiding unnecessary data storage and transfer on a secondary site.

- Since the Time Travel feature adds up storage cost, define the data retention period carefully on selected tables (consider important fact tables that are changing frequently), rather than defining on the database or schema level unless it is very important.

- Use the Snowflake cloning feature to clone objects, which will reduce the compute cost of copying and the storage cost as well.

Best Practices for Compute Resource Optimization

Here are some best practices:

- Define auto suspend and auto resume on all warehouses based on computational usage. (For certain workloads, you may choose to do it manually.)

- Define autoscaling, i.e., a multicluster policy to handle concurrency load.

- Use a standard or economy policy while defining warehouses based on the wait time SLA.

- Monitor the workload to identify nonrequired operations contributing to overall monthly charges.

- Use smaller warehouses for ETL purposes since data load copy commands do not require large computational resources. Consider using larger warehouses for ETL only when complex computations or bulk loading of huge files are being performed.

- You can use multiple warehouses for ETL based on workloads like small warehouses for small load jobs and larger warehouses for heavy data load jobs.

- Larger is not necessarily faster for small, basic queries.

- Analyze warehouse utilization and decide on the perfect size for warehouses based on warehouse utilization and query complexity.

- Enable a query result cache to reuse the existing query result and avoid rerunning queries on the warehouse utilization and query complexity.

- Define a statement timeout in seconds to avoid unwanted long-running queries.

Summary

Snowflake was born from the idea of bringing the capabilities of a traditional data store while at the same time enabling elasticity and scalability of the cloud without worrying about things such as costs, performance, or complexity of managing the system. It offers dynamic, scalable computing power with charges based purely on usage. Snowflake can scale up and down based on need while providing exceptional performance. In this chapter, we summarized the best practices for data loading, query, cost, and compute resource optimization.

References

This chapter contained material from my online blogs, as mentioned here:

- https://soniruchi2891.medium.com/cost-optimization-in-snowflake-75610698ba50

- https://soniruchi2891.medium.com/best-practices-for-data-loading-in-snowflake-f6f9a1462c9b

Index

Printed in the United States
by Baker & Taylor Publisher Services